Third Edition

The Origins of the Tuning System

Fingerprints, Movies and Nature

Discovering a Change in the Magnetic Field

The Sixth Sense

Walter Dickson

Book design by David Russ, Russ Ranch Productions

ISBN 978-0-578-44363-8

10 9 8 7 6 5 4 3

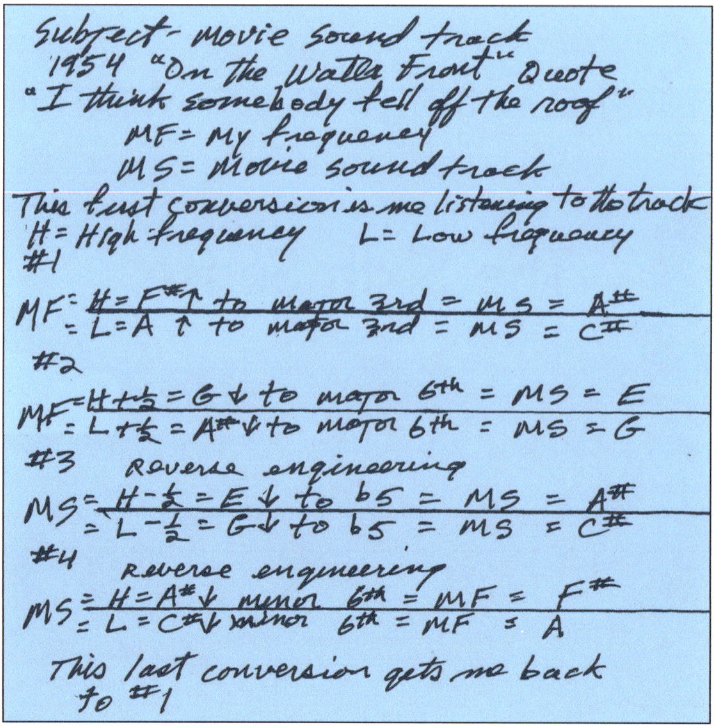

Subject - movie sound track
1954 "On the Water Front" Quote
"I think somebody fell off the roof"
MF = My frequency
MS = Movie sound track
This first conversion is me listening to the track
H = High frequency   L = Low frequency
#1
MF: H = F# ↑ to major 3rd = MS = A#
= L = A ↑ to major 3rd = MS = C#
#2
MF = H + ½ = G ↓ to major 6th = MS = E
= L + ½ = A# ↓ to major 6th = MS = G
#3 Reverse engineering
MS = H - ½ = E ↓ to b5 = MS = A#
= L - ½ = G ↓ to b5 = MS = C#
#4 Reverse engineering
MS = H = A# ↓ minor 6th = MF = F#
= L = C# ↓ minor 6th = MF = A
This last conversion gets me back to #1

Near the end of the second edition of this book, I started working on movie conversions like this. The equation above needs studying and will be the subject of my next book. This concept was written in 2018. Now on 11/30/23, I checked the beginning of On The Waterfront *and it checks out E over G or E/G which becomes E/G/A#/C#. Later in the movie, it becomes D#/F#/A/C. Towards the end, it becomes D/F/G#/B which are all the notes in an octave being assembled by the diminished 3 half step scale. The equation above no longer applies to 2023.*

Preface

Hz (Heinrich Rudolf Hertz) = unit of frequency, equal to one cycle per second, or one vibration per second.

I've always loved the piano and wondered why it's based on middle C. If the ancients were calculating the sounds of flies and other common insects as the key of F#, this would put their low magnetic frequency on the key of A related to 440 Hz in today's standards. Although some people choose to play in slightly lower keys such as 428 Hz for string pieces and other selections of music based on the tuning of the original composer.

Thinking back to the note of F# in the days of old with the horror of diseases and death all around, life was very fragile. The main thing that comes to mind would be flies. Flies were the first insect to arrive at a scene of death and can sense death miles away. They are aggressive in their pursuit of death. Screened windows weren't in use until the 1890s, along with poor ventilation, etc. The fly is a high frequency insect and today in 2018 they are in the key of F or F# varying slightly as they react to the magnetic field.

The other insect of interest is the cricket mainly being a multi-frequency insect. In 2018, crickets are in the key of A and A# varying slightly as they react to the magnetic field. The fly is a daytime insect and the cricket is a nighttime insect—not that night or day changes the magnetic frequency. I'm just selecting these two insects for study as all insects and animals are affected by the magnetic field.[1]

I'm picking these two insects because I know that in the early days of tuning people had direct access to these sounds.

My research has produced different variances in frequency analysis. The search for the tuning system can be traced to early music notation dating back thousands of years to the Greeks and the legendary Chinese Yellow Bell System. All ancient cultures had a type of tuning system and some ancient instruments have been recovered. Apparently, it was the Dark Ages that lost much of this information. My guess is the tuning system is based on nature. If the magnetic frequency I wrote about in pre-2009 and post-2009 is factual, this doesn't mean the magnetic frequency was behaving in this manner back in the Old Times. The ratio of frequencies could have been different from today. The low frequency could have been C and the high frequency could have been F# written down as F#/C, as today's magnetic frequency equals F#/A, but varies (see page 19).

Through the ages the tuning system leaves me thinking the opposite of F# representing death, might be the key of C representing the life cycle, putting the key of C in the center of the piano. The piano being based on the lyre.

If the movie graph on page 31 is correct, the flies and crickets' keys would move with the graph as it indicates the movements of the magnetic field. If the magnetic field changes rapidly how would migrating animals know where to go? I would say it's obvious to them and they know without a second thought. There must be some kind of basic frequencies that no matter what is understood by migrating animals. At this point, I haven't located it. One guess would be it's similar to an x-ray of your hand. The energy passes through your hand and exposes your bones, as in the earth's mantle influencing energy moving outwards through topography. There may be an alignment of energy directing migrating animals.

Claire Bass was a sculptor and told me on her deathbed in San Geronimo, "there is nothing perfect in nature" and "there are no straight lines in nature". This is of interest because our tuning system is fairly exact, except for temperament.[2]

There may have been a time when the low frequency was up at the key of C which is 3 half-steps higher than today. By 2018 standards, if the key of C was the low magnetic frequency, the high frequency would be A which equals A/C. If the low frequency was the key of C for an extended period of time this could have affected the general concept used in the Gregorian Chants. But my feeling is still leaning toward the fly theory relating to death. I believe much of this information came from the Far East. Also, the magnetic frequency may have a part to play with climate change.

The actual blow flies (calliphoridae) that I listened to in the studio were larger and darker than this species. This house fly (musea domestica) is fully grown at 1/2". This smaller fly is one octave higher in pitch but potentially can use the high and low frequency as it navigates being much more tempermental than the larger specie.[3] *(Photo by Frank Lewis)*

l-r Mel Dickson, Walter Dickson, Bill Everett. Walt on his 750 Norton Commando at Dickson Ranch in front of Mel's wood lot, circa 1986. (Photo by Laura Szawarzenski)

Researching the Question

In 1993, Phil Wood loaned me the 1962 edition of the Harvard Dictionary of Music. On page 17 under "Aesthetics of Music" is a quote from R Schumann, "Music, however, is a poor orphan whose father and mother nobody can name. But, perhaps, it is precisely this mystery of her origin which accounts for the charm of her beauty."

I went to the internet to the Harvard Dictionary of Music; Fourth Edition; and looked up tuning. It went from tuba on page 923 to tuning cone on 924, tuning fork, tuning pin, tuning slide, tuning wire and that was it.

Looking up Merriam-Webster's definition of music (a) the science or art of ordering tones or sounds in succession, in combination and in temporal relationships to produce a composition having unity and continuity; (b) vocal, instrumental sounds having rhythm, melody or harmonies.

Reading books of ancient civilizations of China, India, Greece, Egypt and other cultures from around the world, I realized the tuning system not only started way back in time but also varied in pitch and scale.

Today when I'm tuning instruments it's at the standard A=440 Hz.

The books never seemed to suggest or point to the origins of the tuning system. Just the dilemma of tuning old pipe organs back in the 10th century and moving forward in time is a lifetime study.

The European tuning system varied from approximately A=400 Hz to A=456 Hz until France in 1859 decided on A=435 Hz at 15 degress Celsius or 59 degrees Fahrenheit for concert tuning.

Theodore Thomas, a renowned director and conductor working on the East Coast of the United States, decided to install steam heat in concert halls starting in 1883. This increased the temperature up to approximately 68-72 degrees Fahreneit. Warmer temperature raises the pitch of wind instruments bringing the A to 440 Hz. Between 1900-1955 the A= 440 Hz was slowly adopted for international concert tuning.

Search for JC Degan (Percussive Arts Society) for the science of tuning, also mathematical tuning Pythagoras (570–495 BC).

Sensation

As in any science when the door opens on a discovery, this should lead to more expanding discoveries and of course more questions. For myself, this book is like washing an old window that hasn't been looked through in a very long time.

In science, it's so easy to use sight at first thought instead of a hearing type of sense. Inside the inner ear are small crystals called, otoconia.

In my case I have tinnitus and I believe these crystals are picking up high and low frequencies from our magnetic field. I'm now in my sixties and I feel sore and slightly worn out from my physical endeavors. This is a good time for me to write about some musical experiences and find my tuning system.

I'll leave out most of the trauma that along the way tripped me up but also shaped my character to write this short book.

As a Child

I'll take a few minutes and write about what started me thinking about this.

I believe it was 1959 to 1960 and I was about 5 years old. I was home sitting at the breakfast table enjoying mush with my mother, Thelma. Mom had finished breakfast and was cleaning the kitchen doing dishes. I watched her washing as I finished up.

I had recently become aware of a strange sensation in my head. A vibration—a sound. It didn't frighten me, it was just there. I decided to tell Mom about it. I walked over to her as she scrubbed away and said something like, "I hear a sound in my head." She then stopped, dried her hands and said, "What did you say?" She bent down and said again, "What did you say?" and again I said, "I hear a sound in my head." She then took me to the kitchen table, sat me down and began asking me questions. I explained it, no doubt in the way any child would try. She got up went to the telephone and called my Aunt Grace who was my doctor and soon I had an appointment at an ear, nose and throat specialist.

A day or so later we went to San Rafael to the appointment. Once in the office and then in the examination room, a nurse sat me on the edge of a bed with a waxy paper on it. It made a crinkling sound when I moved so I tried to remain still.

In walks a middle-aged man in a long white overcoat.

He asked me a few questions like how do I feel, any accidents, etc. After looking down my ears, he wrote in his notebook, turned to my mother and said something like, "most likely Walt is inflicted with tinnitus. (The hearing of sound when no external sound is present) He goes on, "It's generally something we can't remedy. It could be a life-long issue." Remembering back, I didn't like the man's attitude and that's why I remember the memory so well.

My Mother was a bit upset but after all, she had tinnitus too. So I've been living with a fairly loud type of sound or energy in my ears for all these years.

Growing Up

During my growing up period in the West Marin area, I had the Dickson Ranch, hills, creeks, lakes and more than enough fun with horses, kites, bicycles and all the other animals to play with. I also had lots of young friends from the baby boom-era to run around with.

In the late 1950s and early 1960s, it was easy living for all kinds of musicians and artists because the rents were low. In 1954, one could buy a nice house and property for $1,800.

My Mother was interested in music and started befriending musicians and the next thing I know in 1964 she starts singing opera. Soon after, opera singers, jazz musicians, classical piano players, celloists and the lot gave her the idea to start a monthly music session on the first Sunday of every month from about 1968 to 2010.

I remember sitting under a small grand piano watching and

listening to musicians enjoying themselves. Sometimes my friends would come over and we'd be tearing around like wild animals.

Dickson Ranch, painter Ray Strong, circa 1947

In about 1960, I started taking piano lessons that lasted for a few years, then in 1964 all my friends and I got interested in guitars through the McConnell family. I believe the first guitar player I remember hearing was John Rayley, Pete and Kevin McConnell's uncle. It wasn't long after that in 1965 we had electric guitars and we started a small band with Pete, Kevin and myself.

In 1968 the original Sky Blue Band broke up and we asked if we could have the name and they said yes. Now Sky Blue Band just turned 50 years old. By 1968 the band was growing with teenage players.

Around 1961, the folk music scene was in full bloom and was being played at the McConnell parties usually with guitars. This was part of the California reality and changes were just starting as 10s of millions of dollars were headed toward the Bay Area for recording contracts. The migration had started.

l-r Kevin McConnell, Pete McConnell, Walter Dickson, 2018 (Photo by Rita McConnell)

Rud. Ibach Sohn. This piano, built in 1903, belonged to Fae McNally and is owned by Walter Dickson (Photo by Frank Lewis).

Special thanks to Wypke de Vries for his woodworking expertise.

Moving to Discovery: Fingerprints

Some years ago around 1993, I was recovering from five years of recording which was financially tough and I also took one year off work due to a severe dog bite to the right shoulder. I rented my house and moved into my music studio. I had a comfortable couch outside as I do today and lots of music gear inside. This gave me time to study books on the subject of music which led me to the tuning system. I could read all day and play music all night, although I had a sore shoulder and I was broke.

One rainy stormy night alone in the studio, I was tuning the Ibach grand piano with the help of an electronic tuner. I was trying to understand temperament.

I had a strange sensation in my head and it was getting stronger in intensity. I walked over and opened the door. It was pouring rain with wind and thunder. I stood there as an electrical buzz was humming in my head, and right then, bang, lightening went off right above the studio. A huge blue light lit the area up for a second. I rocked back to the piano and hit middle C, and there it was, the frequency in my head. It wasn't the exact octave but it was the right relative note. I grabbed the tuner and walked it up to C7, very high. At that moment, it seemed to all come together. I briefly slipped into another dimension.

The Snap

I jumped up and yelled, "I've got it!" and snapped my thumb and middle finger together, and as it went click, I noticed it was the pitch of C.

I looked down at my fingertips and took a few seconds to realize I was looking at the magnetic field, or a picture of it. Probably at the moment of conception, the magnetic field is imprinted on our genes or DNA as a snapshot of magnetic time.

At the time I was unaware there were three basic energies to the magnetic field and a mystery frequency. On that stormy night I became aware that there was one frequency, or at the least, I thought there was one frequency. The next day I snapped my fingers and the pitch had changed. I couldn't believe it. The pitch was somewhere else—what happened?

This threw me into a type of depression for a time. I didn't know the magnetic field was capable of changing its frequency. It turns out the magnetic field was changing frequencies, especially before the Mayan calendar ended.

In recording studios and musical situations, hand clapping and finger snapping can be influenced by the key or keys of the song or musical subject. That's why most of the time clapping on recordings are in pitch with the songs root. I call this molecular vibration. The ears tell the brain what key the song is in, figures out the root and electrically vibrates the molecules and cells of the body (hands, finger snap, etc.) to be in pitch with the root of the song.

Description of three basic energies of the magnetic field:

1. The low frequency is approximately 3,521 Hz (A)
2. The high frequency is approximately 5,916 Hz (F#) in stereo as a hissing sensation like a very high violin string.
3. This third frequency is a wondering mid-level energy that moves above the low frequency. It is at approximately 3,951 Hz (B)
4. Mystery Frequency see pages 17 - 18

I'm writing this page on Thursday, June 7, 2018, at the studio and my real-time magnetic frequency is A = low. The low = A is the root. The mids are just above the low sitting two half steps up or one whole step up on B. It's a very light energy, but has the potential of overpowering the low frequency.

The high frequency is F# = Crystal Grit Stereo (CGS). The magnetic field is in a flux, almost liquid as in a phasing, waving motion. There are different energy outputs for the three energies and different combinations.

Here is a graph of the three basic energies:

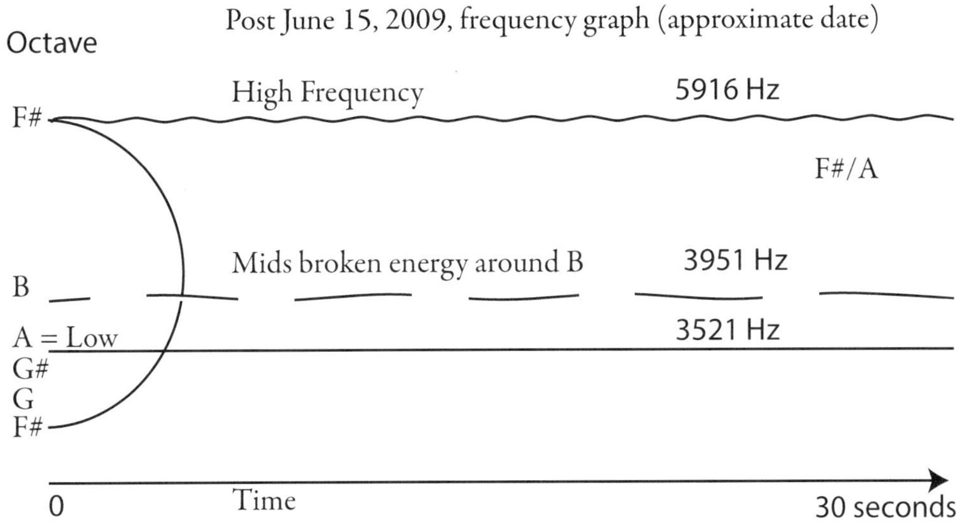

To me, the lines on the previous graph (see page 15) are starting to look like the beginnings of staff lines to written music. Years ago, when I started writing about the tuning system, I always pictured people in the back rooms of buildings discussing the tuning system with various chaotic scenes. I want to mention that the treble clef 𝄞 and bass clef 𝄢 resemble the cochlea and outer ear, respectively.

When I'm charting my real-time magnetic frequency, generally, to find the high frequency I move down three half steps from the low magnetic frequency, then jump up one octave, or down a minor 3rd and up 1 octave (see page 15). The reason I can find the low frequency is that I can hear it loud and clear in my right ear. If moving up from the A equals the low, the F# is a sixth up the major scale. Do=1 Re=2 Mi=3 Fa=4 Sol=5 La=6 Ti=7 Do=8

Occasionally, the high frequency is down one whole step then up 1 octave to the roots dominant seventh. I call this long ratio (LR). This high frequency long ratio becomes a mathematical challenge in the fact that recorded voice frequencies are effected.

Charting Movies

Let's say when listening to a recorded conversation, the low note of the conversation might be F#.[4] This note represents the high frequency of the magnetic field. Take the F# and go up three half steps to A. This note represents the low frequency. I write this F#/A which is similar to the graph on page 15. Everything seems upside-down and it made it a hard discovery. Down is up and up is down. Apparently, the human mind processes the magnetic frequencies in this way as keeping the two frequencies in a closer ratio.

Insects

Photo by Frank Lewis

The mid-frequency of the previous graph is detectable to me only part of the time. It's a very strange, live energy and seems to come and go when it wants. I noticed the ravens and the crows can sense the mid frequencies when I can't. I've spent time with crickets, dogs, horses and many different kinds of birds tracking the magnetic fields frequency which made me aware of many frequency discoveries.

I'll explain the characteristics of the three basic qualities of the magnetic field and some of the insects it affects. Also, I'll examine the mystery frequency.

Horse flies = High frequency
Honey bees = Low frequency
Bumblebees = High frequency (with some low depending on age)
Mosquitoes = High frequency
Wasps (yellow jackets) = Low frequency
Flies = High frequency
Bald-face Hornet = mystery frequency - D, if the low frequency is A
Crickets = Low, High and Mystery frequency

Bald-faced hornet. (Photo by Walter Dickson)

The mystery frequency of D or D# depending seems to be related to the low frequency as a harmonic value. The frequency appears to be the fifth below the root of the A major scale. The reason I'm suggesting this is because of the study of crows, ravens, and insects as the bald-faced hornet. The sounds being generated in this study of the mystery frequency of D, D# or E through the analysis of these animals and insects are always in a lower register than the high and low frequencies.

One insect I've always been interested in is the cricket. I just spent two weeks with the crickets in the Sierra Nevada Mountains. In the evening they start pulsing for a few seconds and then stop. A short while later another cricket starts a different note for a few pulses then stops. Eventually, they find the note they like and begin pulsing. One reason they had such a problem finding their starting note was the root of the magnetic frequency was moving up and down as noted on the handwritten page (see page 19).

There seems to be two different notes that crickets generate. It's the low and high frequency of the magnetic field as in this case on page 147 of my 34th notebook. I'm not sure if the lows are male, female or both. I wrote part of this by the campfire which explains why it is so scribbly. My handwritten page is dated 8/18/18, Saturday evening, one day before the

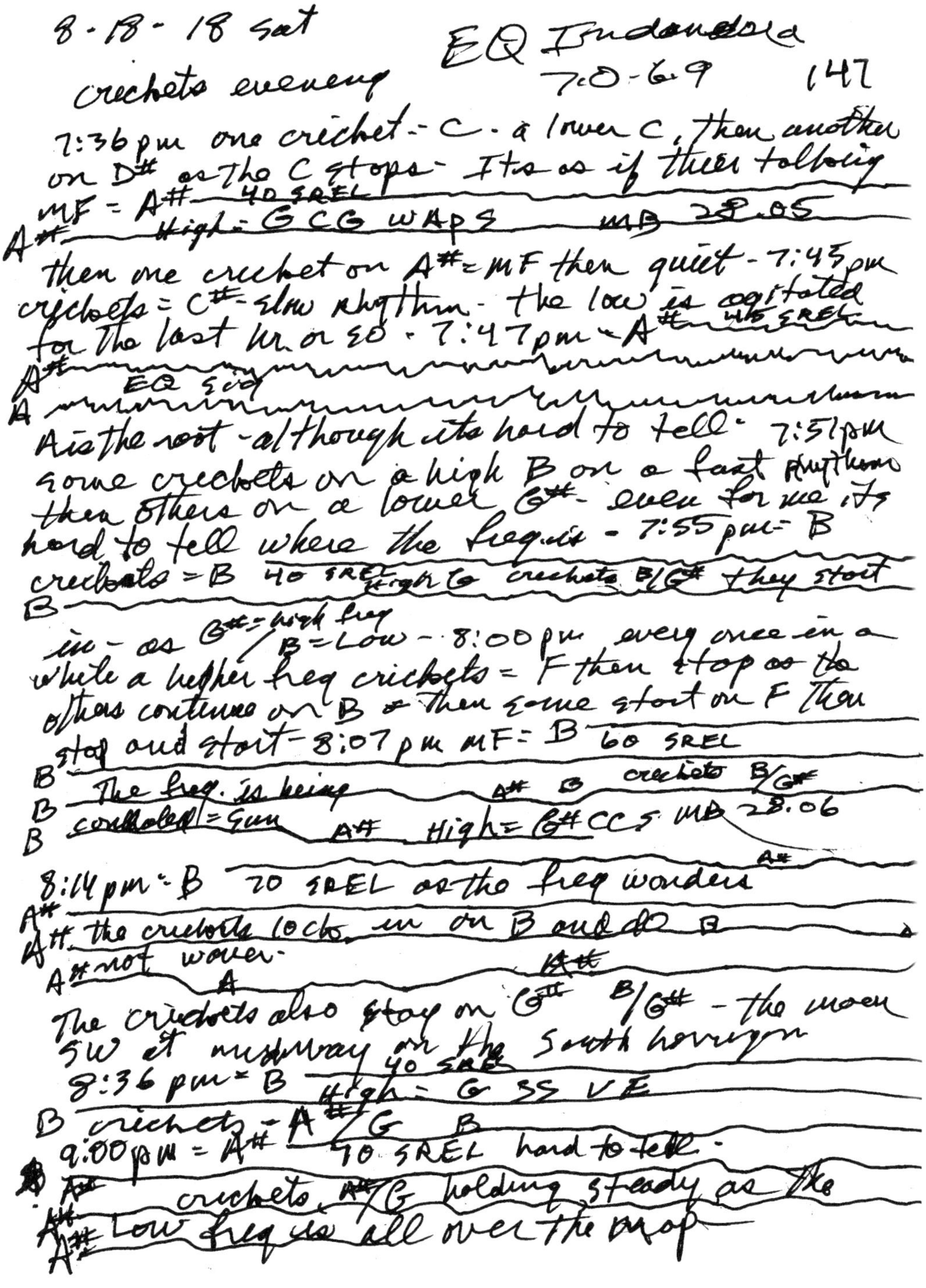

The "EQ Indonesia" at the top of the page I added the next day after I heard about the earthquake on the radio.

6.9 magnitude earthquake in Mataram, Indonesia. When walking around on that night, I could find small patches of crickets pulsing on either the keys of E or D# depending on the real-time magnetic fields frequency. Since I can't hear or sense this frequency, I call it the "mystery frequency".

Here are some of the definitions of abbreviated notations from the previous hand-written page (page 19).

G = high frequency - written as G/A#
A# = low frequency
40 SREL = 40 seconds right ear = low frequency
MB = My barometer
EQ SIG = Earthquake Signature
SS = Stereo Strings
WPS = With Phase Stereo

1. The low frequency of the magnetic field might be produced by the friction of the tectonic plates.

A. Normal low frequency 15 seconds, a slight wave

B. Low with a flutter = mid & low combo

C. Slight agitation creating a pulse

D. Pre-earthquake

E. Earthquake. Blink and aftershocks

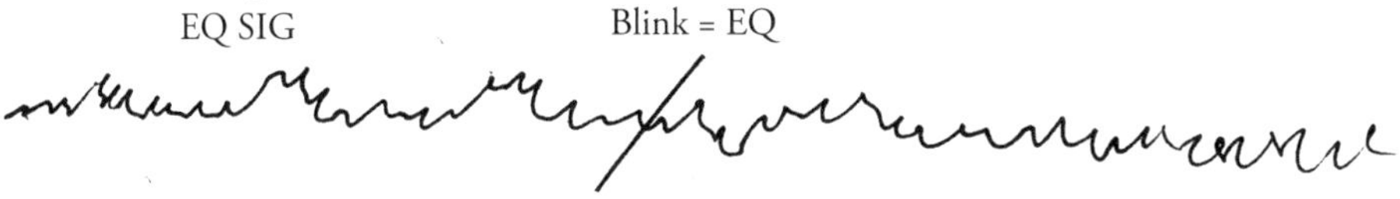

F. Waves. Rolling possibly caused by changes in the Sun's magnetic field

2. Graphs of high frequency

A. Smooth string stereo effect

B. Stereo strings with phasing waveform

Stereo is right ear, left ear varying in energy. The stereo can lean right or left or pan right or left. Most of the time when the high pans, it pans right.

C. Crystal clinkers (CC). Sounds like a crystal chandelier rocking back and forth causing a clinking sound. Each dot represents a quick lightning strike or elecrical pusle.

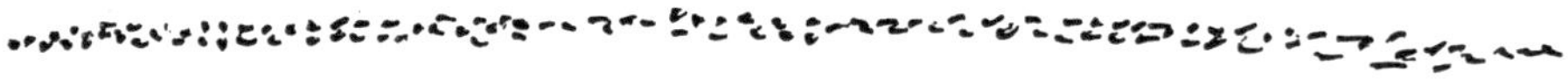

The above graph is approximately 8 pulses per second. Generally in stereo and random, similar to a Geiger counter.

There must be energy generated from liquid magma and internal forces of friction, especially being made of iron.

3. The mids exist above the low frequency so I will include the low in the graph.

A. Low = A

Mids = a broken line of the energy waving around B

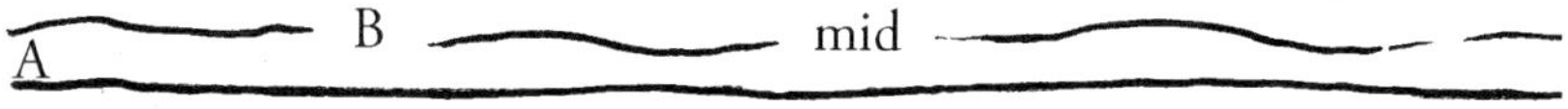

B. Low = A

Pops of energy that vary in length. These dashes represent longer lightning pulses.

C. Low = A
Mids are unbroken fluttering energy. This energy sounds like a bat flapping its wings in flight.

D. Low= A
The mids are aggressive and attacking the low.

E. Low = A

The mids attack and take the place of the low then climb back up, in this case to either the key of B or C. Generally after 3 to 10 minutes the mids either dissipates or are absorbed by the low frequency or vice versa. Sometimes the mids have enough energy to pull the low frequency up one whole step, then the low and the mids combine into one energy. There is a strong possibility this attacking mid frequency could be a frequency from the sun.

The previous graphs are not unusual because some of the energy I'm sensing is associated with Japan and the surrounding areas (see page 19). I'm suggesting the association to the areas of Japan because of predicting earthquakes in the area. There is a possibility that the Sun's magnetic energy could distort the Earth's magnetic field enough to create some types of earthquakes.

Here's a story on what helped me find the high frequency.

When I'd be studying in the studio there would always be one or two rather large black flies buzzing by my head. At first, it was annoying as the mosquitos were. One question I asked myself was where do these insects get their information? I started listening to everything in the context of music and by doing so realized these insects had a logical frequency. In the flies case, it was F# if the low magnetic field note was A. I was able to identify the flies vibrating wings at approximately 1480 Hz equals F# as the relative note of the high frequency pitch which equals F# at 5916 Hz, four octaves up.

Energy Changing Frequencies

I listened to television shows from New York with the word "live" on the screen. There may be a slight delay in the broadcast time of 7 seconds due to censoring. With my tuning machine, I could identify when a pitch drop occurred from the East coast and count one thousand one, one thousand two, etc. After approximately 11 seconds the magnetic pitch frequency drop would affect where I was in California. So let's say we have 18 seconds for the frequency change to travel 2,500 miles from New York to California. This means the energy traveling clockwise or traveling east to west is moving approximately 138 miles per second on the planet's surface. This is the slowest I have calculated this energy and generally, the speed is faster.[5]

I've been thinking about how this could happen. The only explanation I can come up with is the following. Looking down at the solar system from the north, the sun spins counter-clockwise at approximately 4,400 miles per hour at the equator and casts a slightly curved magnetic energy field 93 million miles to Earth and beyond. The slightly curved energy arcs away from the direction of the Sun's rotation and revolves counterclockwise at a higher speed than Earth's counterclockwise revolution looking down from the north.

The Sun's magnetic energy field contacts the right-side or east-side of Earth's magnetic field first, bringing information causing a clockwise transfer of energy (see page 44) that moves towards me when I'm at the AM position on earth spinning counterclockwise.

The Sun's magnetic energy envelops earth on the PM side from east to west and meets at the approximate location of 2–4 am at the leading edge of earth's travel around the sun. The Sun's magnetic energy is like a tape from a tape recorder contacting a playback head being Earth's magnetic field.

The Solar System

What it looks like to me looking down from the north.

In my mind, I'm reducing the sun and planets down to the size of marbles. It appears the sun and most of the planets are rolling on the right-hand side or the outside of energy field rings causing the counterclockwise rotation and revolutions.

Venus appears to be riding on the inside of an energy ring causing a clockwise spin and Uranus appears to be riding in an energy wave rolling along like a ball.

The sun rolls along to the right on the outside of an energy field, rolling down and creating a very large helix at a precise speed. This falling effect could be caused by energy flowing around our system from an expanding galaxy.

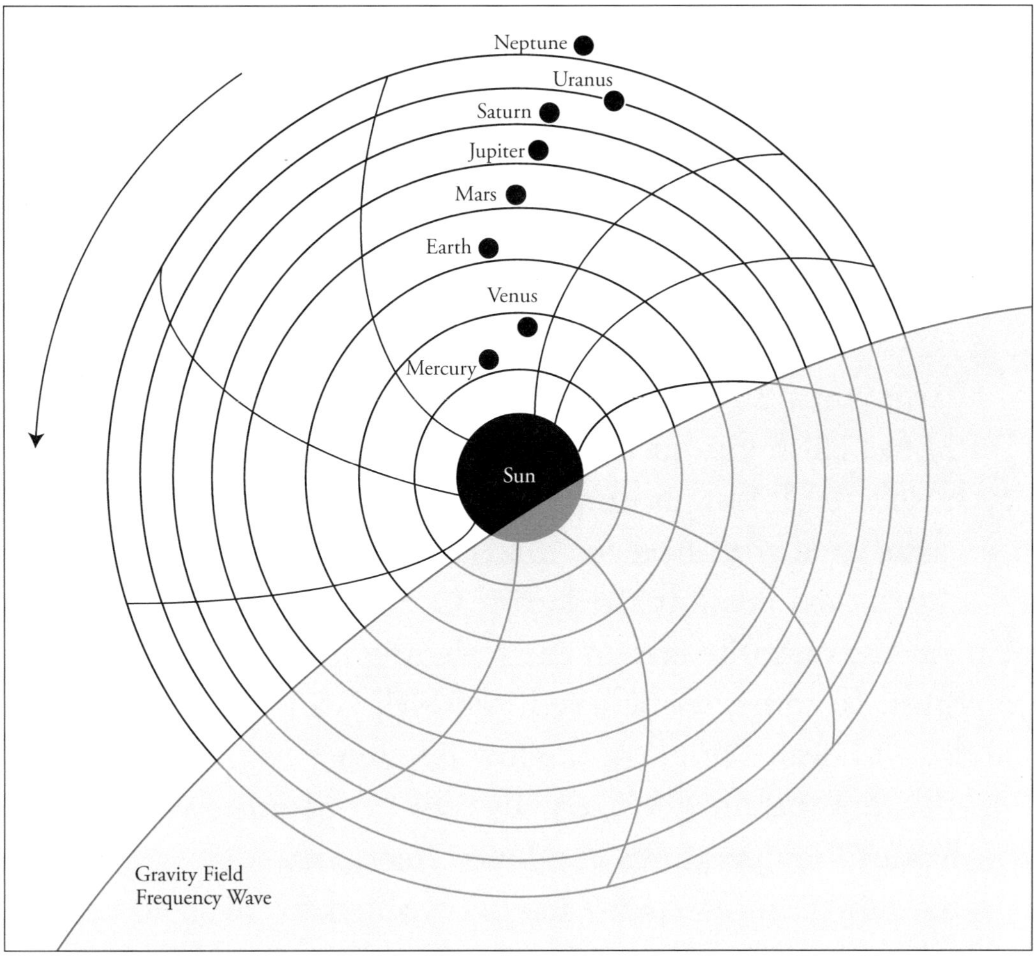

View from North looking down on the Solar System (Graph by David Russ)

Our system is located on the edge of a spiral arm and is moving as water spinning down a drain, as if either on one side or the other of a huge equatorial divide, suggesting the universe is round and spinning.

I imagine the arm that leads down to the center of the Milky Way galaxy is in a state of falling toward the center. The center might be moving away from us and in that case, energy could be blowing by us most likely in a spiral and creating the effect of falling but still being pulled toward the center, as in a push-pull effect.

Pieces of the Puzzle

I've been working on hints left behind from the Hebrews, Christians, Buddhists, Hindus, Islam, Mayan, and other cultures. The Gregorian chants suggest that the key of C is the sound of God as other cultures suggest. Other cultures also suggest the key of F# is the sound of the Devil.

Hieronymus Bosch painted scenes from the last supper (1482) to hellish visions of chaos and fire—to use the word psychedelic is almost an understatement. At any rate, Hell has been associated with heat for thousands of years. If the key of F# = heat, then possibly the key of C, at a higher frequency, being at the opposite end of the scale might mean cooler temperatures. To some, it's believed that Hell is a place of perpetual fire and heat. I'm using the low frequency as the example, low=the root meaning A/C equals cool temperature vs. D#/F# equals warm temperatures at a lower frequency.

This subject interests me having worked for over 50 years in the Heating Ventilation and Air Conditioning (HVAC) business. The low frequency note seems to have the most influence when studying differentials of heating and cooling.

Movies Past and Present

As I began to figure out the magnetic frequency in real time, I also wanted to figure out a way to trace it back in time. The thought was in my mind for 13 or so years—and as I started to generally figure out the magnetic field, I thought of this:

What if the human conversation is an exact print or extension of the magnetic field. This would be due to the energy being transferred through our bodies at an exact moment. After all we're living in the middle of a force field.

If this is true I should be able to listen to movies and hear changes to detect the magnetic field. I started listening to human recorded conversation patterns in movies and speeches. During scenes of conversation, the pitch would drop most of the time. Some scenes stayed the same for quite a while like in 1929, *Broadway Melody* 1 hour 7 minutes into the movie the frequency pitch stays around D/F to C/D# for quite a while. C = high frequency and D# = Low frequency.

Listening through and charting old movies usually consisted of noticing frequency drops of half-step increments. When I'm charting movies I'm studying from scene to scene.

One scene might start on A then go to G# – G – F# and so on. The next scene being filmed possibly at a different time and location might start on D then drop to C# - C - B over a time period, or even stay the same frequency as it started. The combinations in this study are becoming more complicated.

Just like my finger snap changing, so did movies and taped conversation. I began charting TV shows, movies and anything recorded. I would write down the release dates and somewhat calculate when the scenes were shot and recorded.

Then I came across a Global Land-Ocean Temperature Index graph and noticed the jagged lines representing temperature changes during time periods. These up and down lines didn't look right, although they were generally headed in the direction of rising temperatures. I wondered if the magnetic field could be involved in these jagged lines.

When charting movies from 1929 to 2019, there were differences in pitch frequencies. The one major difference of importance is the movies filmed before and after the approximate date of June 15, 2009. Before this time period, movies drop by half steps in pitch during scenes. After this time period, movies generally stay in the keys of G/A#, F#/A or F/G#. This can be proved by listening to movies and finding shooting schedules such as "The Karate Kid" release date of June 11, 2010. One half of the movie is the old frequency, as an example, C/D#, D#/C, E/C#, etc. The latter part of the movie is G/A#, F#/A, F/G# which equals the new magnetic frequency era. During this movie, there are differences in scenes filmed at different locations and times. It seems to me the new era of the magnetic field coincides with the end of the Mayan calendar.

Calculating human recorded speech also depends on the frequency of the magnetic field at the time a movie is being charted.

In the first half of the movie, *Freakonomics* (release date April 30, 2010), the interviews = E/G, D#/F, D/F, C#/E. The second half of the movie approximately equates to F#/A, F/G#, G/A# if my real-time frequency is in the key of F#/A. The last interviews in *Freakonomics* equate to E/G, D#/F# which in every movie filmed from around early June to mid August 2009 appears to be a transitional frequency between the old magnetic frequency era and the new magnetic frequency era. Another movie, *Legion* (release date January 22, 2010), is in the old frequency except for the very end which appears to have been filmed after early June to mid August 2009.

If a movie is one half the way finished, I might start charting it—if it's of interest. One-half a movie is better than none. Try the study of *On the Waterfront*, earliest release date is June 22, 1954, representing a cooling period.

After writing 42 notebooks on movies and real-time magnetic frequencies, I only need a piece of the movie for analysis. Although having the complete chart is an accomplishment.

Another movie I studied, *The Hundred-Foot Journey* (release date August 4, 2014), is of the new period and looks like this: F#/A, F/G#, G/A#.

Quite some time ago I saw a movie that I had charted earlier, named *Ace in the Hole* (release date June 14, 1951). I opened one of my older chart books and followed the movie soundtrack. To my surprise, many of the scenes were a little off pitch and some were way off pitch.

The major problem I had discovered was I hadn't written down my real-time magnetic frequency before I charted the movie. I also found that during a charting session if my real-time magnetic frequency moves, the whole recorded conversation is effected by its change. So I learned my real-time magnetic frequency changes recorded conversations. This is during and post Mayan time. The reason I am mentioning or suggesting the Mayan Calendar is that the coincidence of the magnetic frequency going into a new era is something to be noted. Before getting into this analysis, this next graph will show the old magnetic frequencies and its movements.

The two graphs on the next page are of the Mayan calendar time period that ended approximately June 15, 2009.

The top graph of the Mayan calendar time period indicates a downward diagonal hypotenuse line indicating a normal heating and cooling cycle. The graphs on page 31 may be interpreted as more heat than cool. The vertical frequency line going up represents the low frequency phasing up one octave. Most of the time, the low frequency dictates the pattern of the other frequencies. The stepped line represents frequency drops by half steps. The vertical line on the left represents pitch frequencies.

For this study, one can listen to the recorded speech of Martin Luther King, Jr., on April 3, 1968, in Memphis, TN, "I have been to the mountaintop." On June 5, 2018, my frequency was F#/A. I then charted his speech of 43 minutes 15 seconds long = F/G#, E/G, then holds D#/F# for an extended period – then to D/F and ends on C#/E descending all the way, as time slows down.

In notebook number 34, charting the Martin Luther King, Jr., *I Have a Dream* speech on 8/28/63, I started charting the recording on Friday, July 26, 2018. This session led me to repeated charting sessions of the same speech. The speech is 17 minutes, 28 seconds long. Many times I've charted the speech and came up with different results. I finally realized that my magnetic frequency at the time of the charting needs to be closely monitored to get a reliable take of the speech. At first, I was confused by thinking his top note of C# was the root. This was because my low frequency drifted up from A to A# raising his root to C#. When my low frequency was the key of A throughout his speech, it holds Martin Luther King Jr. in the key A#/C all the

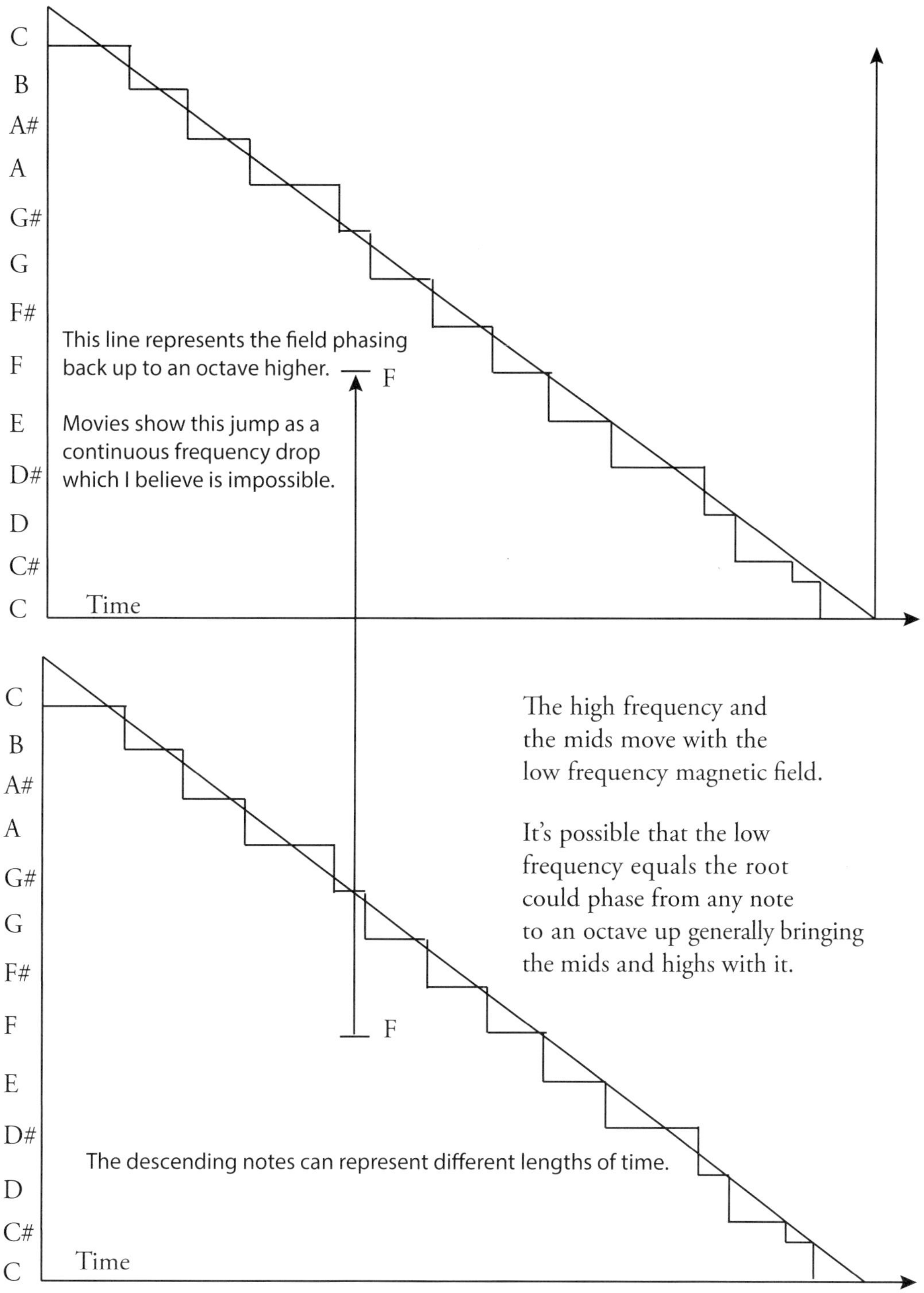

Graph of the old frequency sign wave = one octave

way through. I started checking out the key of the song, *We Shall Overcome*, at the beginning of the film. The key of the song is the key of D no matter what my real-time magnetic frequency was, or is, during the time of the charting.

The song, *We Shall Overcome*, in the key of D, means the magnetic field does not change the key of basic music. The magnetic field seems to sense music as it's own life pattern or entity. Instead of the magnetic field influencing the song, it seems to relax and in a sense, listen. I'm imagining that the magnetic field, as an electrical force field, is causing the atmosphere and all objects to vibrate on a molecular level at the frequency dictated by the energy field. The atmosphere is needed as a medium for sound. This being so means the vibration of the molecules is affecting the pitch of the speech recorded much earlier being projected on a speaker system. The vibration of the atmosphere from the magnetic field moves the speech in directions which may be different from the original recording. Also when looking at the Global Land-Ocean Temperature Index for the time period of August 28, 1963, I noticed a sharp decline in global temperature between 1963 and 1964, one of the largest declines in temperature on the graph. I am suggesting the key frequency of the magnetic field may play a part in global warming and cooling.

In Western music, Pythagoras seems to stand alone in the connection of music tuning to mathematics. Although when reading about this slightly mysterious figure, there doesn't seem to be much information about the frequency of the notes he used as compared to today's A=440 Hz.

The graphs I used are the right-angle triangle, which shows

the hypotenuse as a line that might interpret a measurement of heat. These are basic thoughts of energy measurement output as thermodynamics. The vertical line of the right angle triangle represents frequency. The bottom line of the right-angle triangle represents time. The time varies as the old frequency graph could be short or long as the frequency falls at different speeds and ratios. I am suggesting if the graph stays longer on the high frequency scale this might be interpreted as a cooling effect and the reverse applies to the graph that shows an extended lower frequency period of time. These graphs represent the time period before the approximate date of June 15, 2009. My guess would be a higher frequency force field would absorb cold energy from space and also be tighter to repel the Sun's energy as less porous.

Many questions arise in my mind such as microwaves heating water to the energy used in electric bass guitar amplification systems. I played a gig with my electric guitar being amplified by a 50-watt Mesa Boogie amplifier. The bass, one octave lower, used a 300-watt amplifier and had a hard time keeping with the decibel output of the guitar amplifier even with twice the surface area of speakers. The bass's amplification system was weak in the low frequency.

The ratio of frequencies changes the dynamics of the magnetic field in many ways. Thermodynamics is a word that explains part of these evaluations, as the magnetic field seems to flow and wave like liquid energy. The Pythagorean Theorem could possibly be modified to look like A^2 (frequency-vertical line), B^2 (time), C^2 (hypotenuse is normal heating and cooling), D (frequency track line (stepped line)) = magnetic

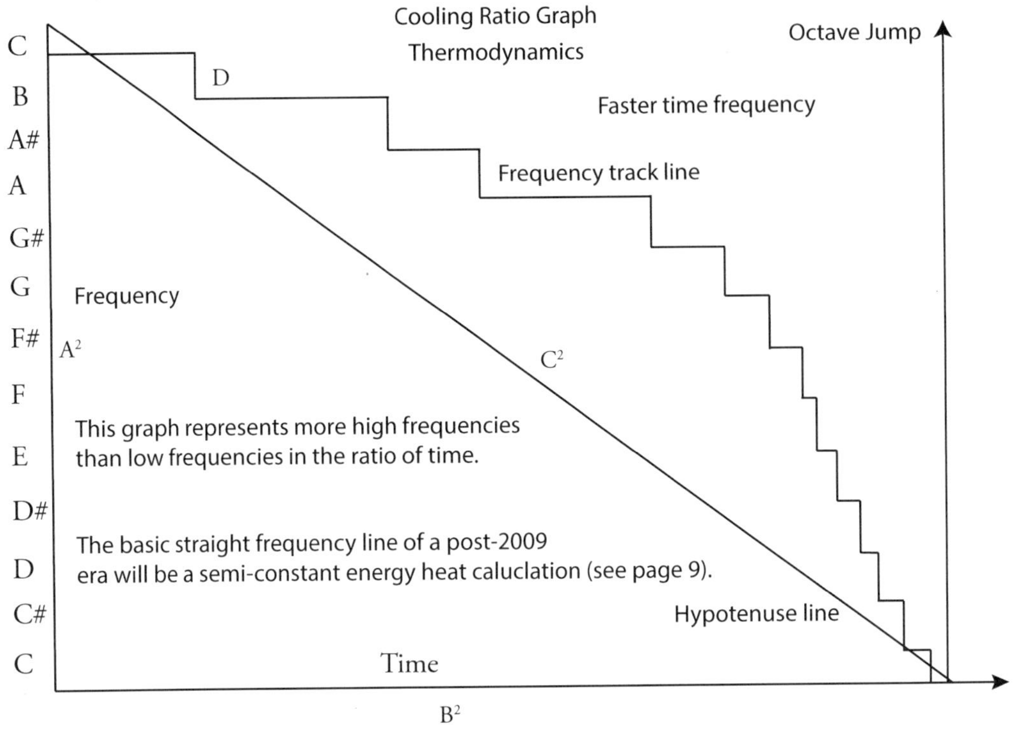
Cooling Ratio Graph
Thermodynamics
Octave Jump
C
B
A#
A
G#
G
F#
F
E
D#
D
C#
C
D
Faster time frequency
Frequency track line
Frequency
A²
C²
This graph represents more high frequencies than low frequencies in the ratio of time.
The basic straight frequency line of a post-2009 era will be a semi-constant energy heat caluclation (see page 9).
Hypotenuse line
Time
B²

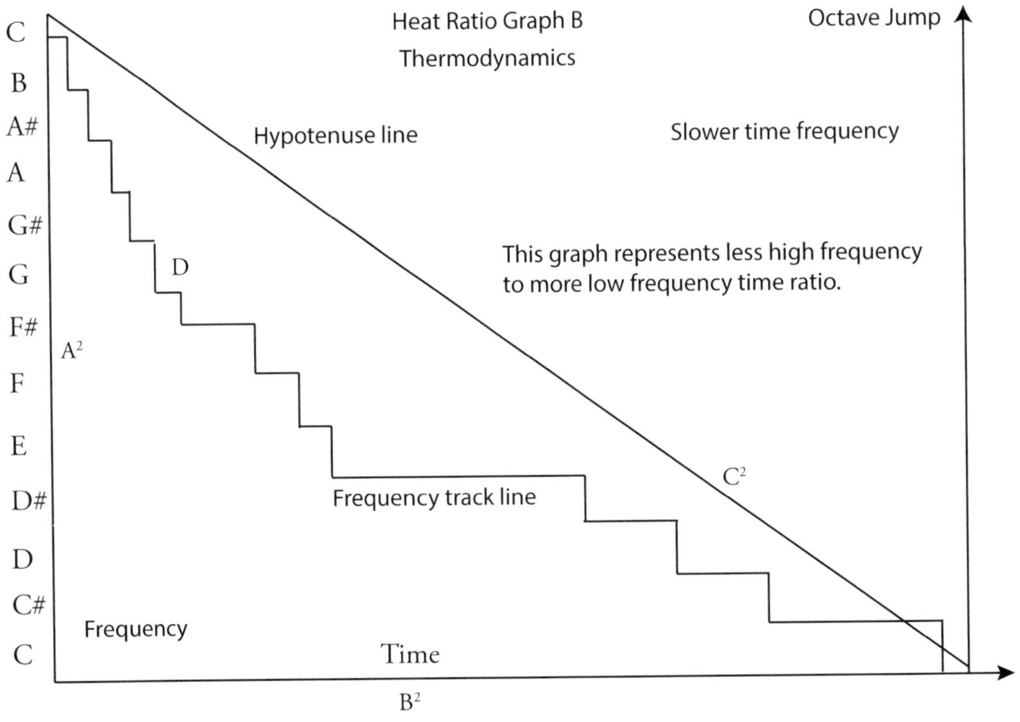
Heat Ratio Graph B
Thermodynamics
Octave Jump
C
B
A#
A
G#
G
F#
F
E
D#
D
C#
C
Hypotenuse line
Slower time frequency
D
This graph represents less high frequency to more low frequency time ratio.
A²
C²
Frequency track line
Frequency
Time
B²

field thermodynamics. With these 4 basic pieces of information, we should be able to calculate heating and cooling differentials of earth plus the effect of space, temperature and the magnetic field. Written music already has the potential of combinations of all the graphs in this book. Written music has vibrations per second, time, and the time value of the notes which could equal magnetic frequency notation.

The right-angle triangle is the only graph that seems to fit the frequency movements of the Mayan time period.

I'm thinking back thousands of years and realizing the ancient cultures had intelligent people working on mathematical equations. These people were studying music and felt it was part of life and it needed to be explained even though it couldn't be seen. It was and is a mystery.

To bring mathematics, architecture, and music together, all one had to do was draw the graph (see page 31) and combine a right, right-angle triangle together with a left, right-angle triangle creating one side of a pyramid. This might explain the pyramids across the world as only the steps and angles of the pyramids could represent increments of descending notes. Viewed from bottom to top, it could be ascending notes. Personally, I'm thinking of descending notes noted in pre-2009 movies.

It seems it would be simple to design an application to identify high low differentials in human speech. This could lead to a magnetic analysis of an individual which might be a readout of Earth's magnetic field. Later our fingerprints could be used for magnetic diagnosis.

I would think crystals could be used to identify the magnetic frequencies, after all, this is the way I detect the magnetic field.

The crystals in our ears are made of calcium carbonate.

Imagine being a person from another part of the galaxy and coming into the Earth's magnetic range pre-2009. They would detect the magnetic sine wave (see page 31) possibly as a distress signal especially if sped up. It would sound similar to a siren.

Back to Movies and Time

If I'm going to chart a scene from a movie, first I choose a talking scene of about one to three minutes long and pause it. I've already made sure the scene is not influenced by music or other interferences of sound, although many sounds are overdubbed later on during the Foley process. This way the scene is clean. I write down the time, then a short graph of 30 seconds of the magnetic field and begin the chart of the conversation.

After I'm done charting, I pause the movie, write down the time and write down another magnetic graph. If the new graph has the same high and low frequency as the old graph, I'm almost guaranteed the movie will replay at the same pitch. If either the real-time high or low frequency changes, the movie conversation could divide into mathematical sequences such as the conversions on the back of the title page. Why? My deduction is the real-time magnetic field recognizes itself through the recorded energy, in this case, human conversation. The magnetic field influences the recording in real-time pitch frequencies—slowing and speeding up time. Sounds far-out, but there it is. Time as a frequency also affects the pitch of clocks and other objects. This discovery means time may be controllable as a frequency.

Moving On

My thought of hearing energy slow down and speed up as a tape of magnetic information does in a studio brought back memories—sometimes not good ones. Different tape machines can play at different speeds if the calibration wasn't set the same. It's the difference of speed, pitch and time. What if there's a different time that's separate from our mechanical cosmic time movement? Light and similar energies are calculated at 186,000 miles a second. The magnetic field may be different, being a more liquid type of energy, oscillating and having human traits–or humans having magnetic field traits.

Obviously, the magnetic field protects us from harmful energy and has an influence on our health and aging process. A note: In my family's bathroom, the walls are covered in veneered plywood having a beautiful pattern of what looks like a picture of the magnetic field as seen on the cover of this book.

The low frequency, being the root, seems to influence what the other three energies are doing. So, if the low frequency is influenced by the tectonic plates slowing down to create the old frequency drop, the tectonic plates would have to slip to create a frequency of an octave jump up as the graph on page 31 shows. The new era of the magnetic field is generally straight in normal mode as the graph on page 15 shows.

But this doesn't sound right.

Thinking Evidence

What would make the magnetic frequency change from the old to the new?

1. Earth rotation – it seems the same
2. Tectonic plates going to a smooth frequency (not plausible)
3. Sun's magnetic field – to me, this sounds like the answer.

I thought about the friction of magma and it's too much of a change to dramatically match the graph on page 31. If the magnetic field's energy is capable of influencing pitch change during a recorded movie or speech, the magnetic energy from the sun has the potential to influence Earth's magnetic field and all the other planets also. The movie frequency pitch experiment is the only key piece of evidence I have relating to the Sun's energy potential. It seems to me the variance in time is evident. If this is true, the magnetic frequency drop on page 31 not only explains this planet's cycling but possibly explains the Sun's cycling. The one larger factor may be the cycling of our galaxy affecting the sun and the earth. It could be the new and old graphs reflect the energy transfer from the galaxy to our solar system.

The Sun's magnetic field and energy field eventually reaches the heliopause area in space repelling and absorbing the energy from the interstellar region. This energy could easily affect the Sun's magnetic field as the front cover of this book's elongated photo suggests.

Our magnetic field is similar to an electric guitar's pick-up system in the reaction to a string vibrating above a magnetic coil that creates an electric signal creating a note. When the Sun's magnetic energy interacts with the Earth's magnetic field this creates energy changes.

How long has the old sine wave from page 31 been going on? Maybe the Mayans and other cultures can supply part or all of the answer.

Physiology

I've wondered if some of my mood swings are affected by the magnetic field, as in physiology. I know without a doubt, my personality is affected, especially in pre-earthquake frequencies. When an agitated frequency was or is happening, the energy transfers to my electrical impulses and I get edgy and irritable. Once aware of my condition, I learned to deal with it, although sometimes I'm up all-night during pre-earthquake frequencies, as noted in this book.

If I examine the five basic senses—sight, hearing, taste, smell, and touch, they all break down to electrical types of frequencies. Reality is related to frequency. It seems to me if we remove all electrical frequency from reality, there is no reality. There is no life as we know it.

During this learning period, I could hear dogs in the neighborhood express themselves by barking in frustration over magnetic spikes and changes in magnetic frequencies. I realized it wasn't only me that had a magnetic problem. Once I realized I had a connection to animals, I knew it was time to calm down and get into a listening state of mind. The simplest way to analyze this was to take notes which ended up in multiple books.

Differences in seconds, minutes, hours, days, weeks, months, then years. As the field of discovery grew, I was able to keep day-to-day journals.

Once breaking the movie code, I was able to go back in time and study the past magnetic frequency from the point of what my reality was at the moment. What I learned was research never stops. It's about following your thoughts to the end, or at least toward the end, knowing there is no end.

First off without long-time friends, none of this would have happened. It's not the pages of this book, it's the 8,500 pages of notes that support this book. Then working to condense the notes down to a summary.

Frustration in the mind becomes an example of an obstacle in your life that needs to be explained or studied in a way that brings comfort and rewards.

In thinking of the challenge of understanding gravity, the more I think about it, the more evidence is gathering in my mind that gravity is related to frequency, as the more matter that's available, the more frequency that is created adding to the energy of gravity. Gravity equals mass equals molecular frequency.

To me it becomes obvious that the choices relating to the tuning system and the mysteries of ancient thinking come together in a single but complex understanding of the magnetic field and its relationship to the origins of the tuning system.

The next section - Tree Knots & Veneer - six categories:

1. South looking up at earth
2. North looking down at earth
3. Looking at earth from the equatorial plane
4. Knots representing the sun
5. Moon knots
6. The remaining 8% of knots are from many different categories

Study of Veneer, Knots and the Magnetic Frequency

This added section of photos of veneered plywood paneling and knots of other wood samples are an extension of the work I've done on the study of fingerprints and the magnetic field. As I study these pictures, I imagine that the frequency lines represent different variations of vibrating electrical frequencies.

When a seed is germinated into a new life form, it's most likely at that millisecond time period that a magnetic photo of electric energy is imprinted on its genetic plans. This magnetic snapshot would give the tree the latest update on solar and earth's energy conditions for survival.

The photo above of a knot in veneered plywood paneling represents earth' and its magnetic field. All following knot photos by Frank Lewis

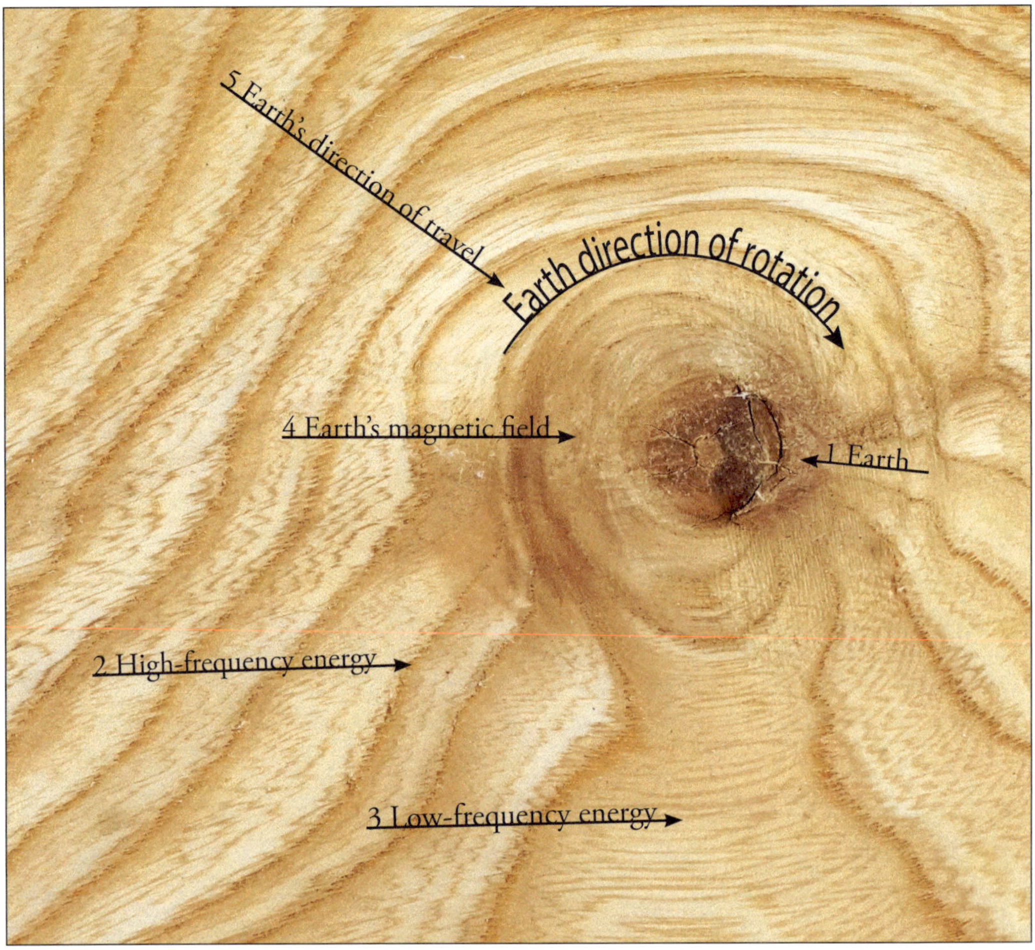

This photo has five different characters. 1) Earth; 2) the lines equals high-frequency energy; 3) the space between the lines equals low-frequency energy; 4) Earth's magnetic field; 5) Earth's directionl of travel.

Looking at a wider angle of the previous photograph, we see the sun's magnetic field being compressed on the trailing edge of the earth's magnetic field (left side of photo). This is the result of the sun spinning at approximately 4400 miles per hour and casting its magnetic field out to earth faster than the earth's revolution around the sun. The perspective of this knot is out in space looking up from the south.

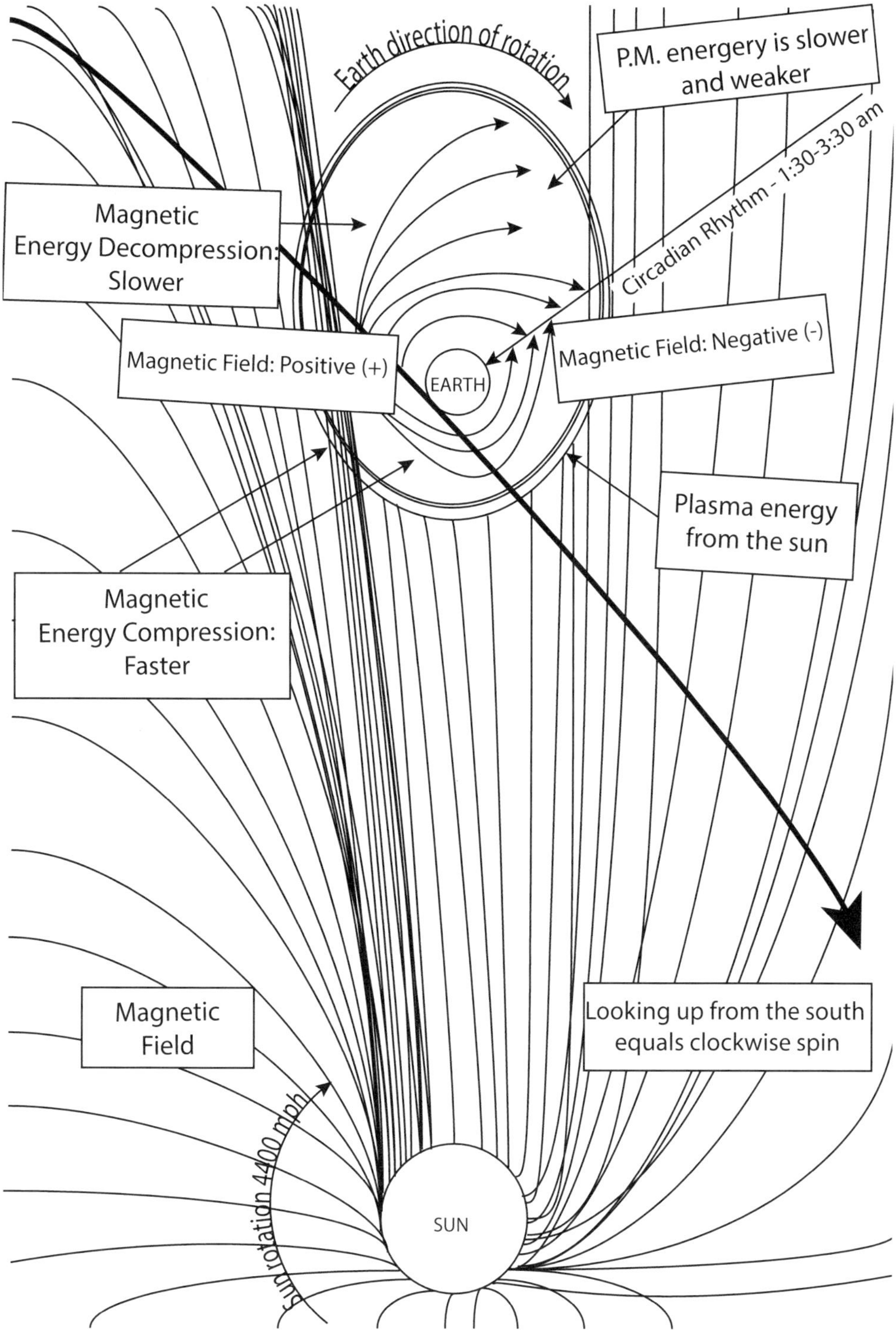

This graph of the magnetic fields are from the point of view from the south looking up. It's the same perspective as the knot on the previous page.

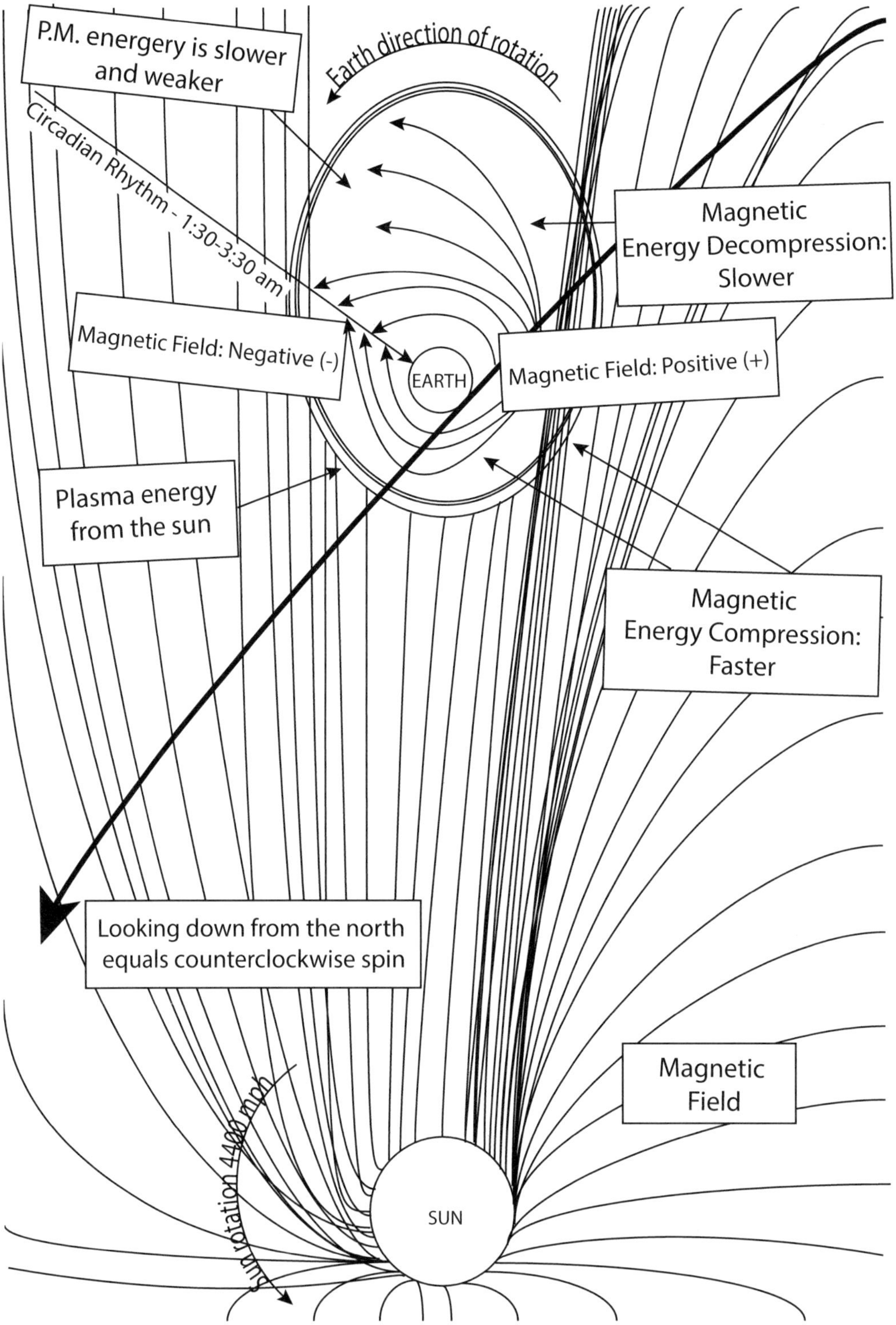
P.M. energery is slower and weaker
Earth direction of rotation
Circadian Rhythm - 1:30-3:30 am
Magnetic Energy Decompression: Slower
Magnetic Field: Negative (-)
EARTH
Magnetic Field: Positive (+)
Plasma energy from the sun
Magnetic Energy Compression: Faster
Looking down from the north equals counterclockwise spin
Magnetic Field
Sun rotation 4400 mph
SUN

This photograph of veneer shows two knots spinning in opposite directions. The knot on the left is spinning clockwise as the graph on page 43. The knot on the right is spinning counterclockwise as the graph shows on page 44. These knots, being from one single piece of wood, means that the seed is capable at germination of picking up both north and south views of the planet from outer space.

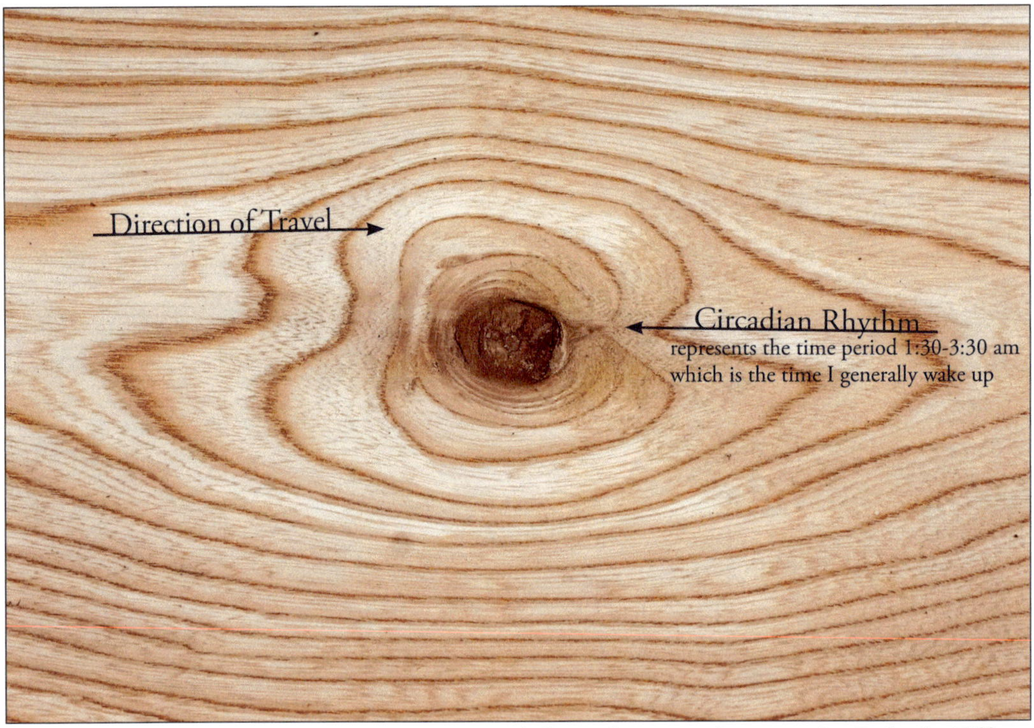

This photo is during the Splash Effect

This knot represents a picture of earth seen from the approximate location of the moon in the equatorial plane. We have compression of earth's magnetic field on the left and the decompression on the right being the direction of travel around the sun. The earth rotates right to left and the approximation of the 3 am location of the magnetic field is on the righthand side of earth. This view or perspective is looking in or out of the plane of the equatorial travel of earth's orbit around the sun. This produces the view of straight lines with a type of hole being forged as earth plunges through an almost static magnetic energy field that drifts broadside through our solar system.

The reason the magnetic frequency lines are almost butterfly in shape is because the MRI or CT scan effect of the sliced cut-away view. The picture, on the right, is what I call the *Splash Effect*.

This veneer shows a 4-foot wide view of a flattened version of the *Splash Effect*. As our solar system moves through space, there appears to be a liquid-like, expanding magnetic energy that looks to be almost static. As our solar system moves through space, it creates this *Splash Effect*. This energy causes the solar system to slant as we travel through space. As the energy travels through our system and we travel through it, the sun and the planets act like rocks striking water causing continual shock waves of energy. The slight differential of the plane of the planets might be the result of this effect. Notice the repetitive patterns of every other energy field being slightly different causing different depths as an MRI or CT scan would be. This log may have been about 28" wide when veneered and shows the same *Splash Effect* on the same side of the tree at a different depth.

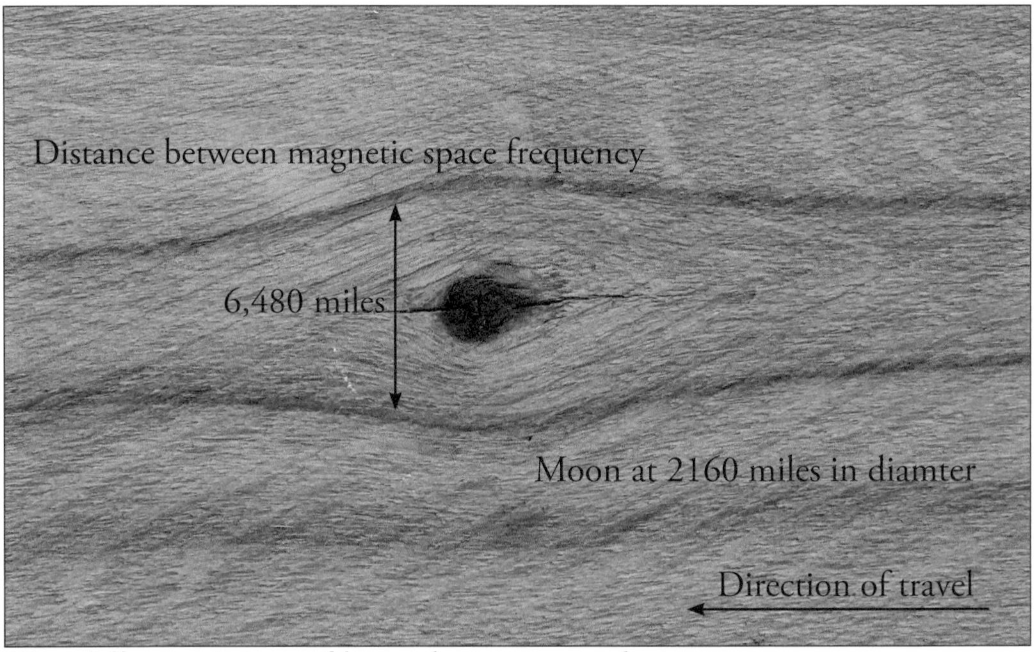

Two different pictures of knots that represent the moon

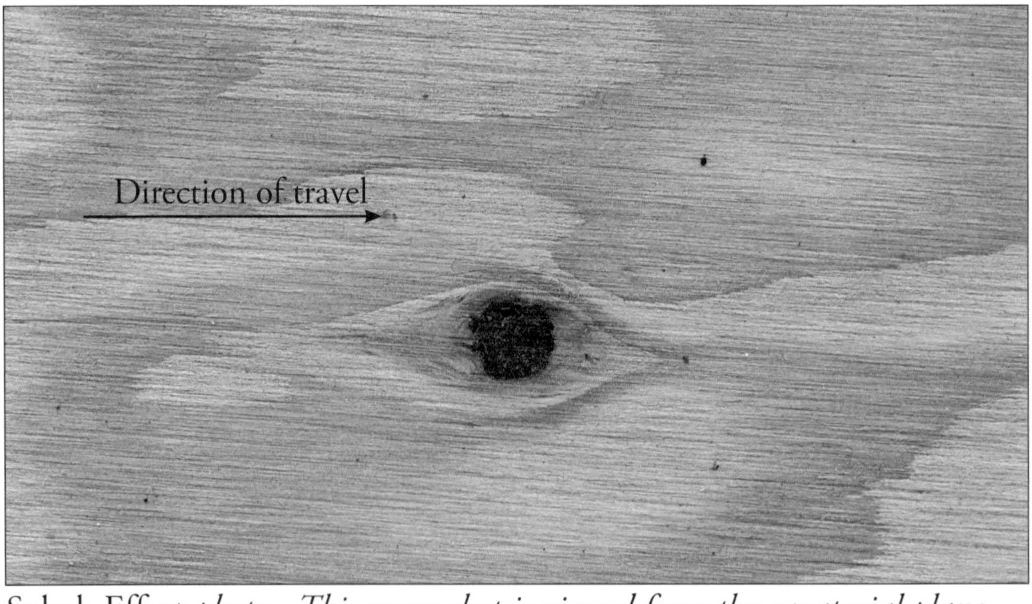

Splash Effect *photo - This moon shot is viewed from the equatorial plane.*

The two photographs on page 48 show another feature of what I believe are interpretations of the moon being from two different kinds of wood. These knots are very small at about 3/4" in diameter. When colorized by computer, there is no evidence of them having a magnetic field. The other thing I see is a faint distortion behind the knots as they appear to be traveling through an energy field. I want to mention all of these pictures have perspective. For instance, the moon is 2160 miles in diameter. This gives us the dimensions of the space frequency energy fields. This also applies to all of the other pictures as perspective or scale.

I've noticed over and over the similarities to the effects of energy fields being represented as a liquid in all the photos being studied.

These pictures represent two different electrical energies that are able to co-exist as two different frequencies, in two different dimensions. The Splash Effect photos indicate the existence of a type of static electric energy that may permeate the solar system and galaxy controlling the frequency of the sun.

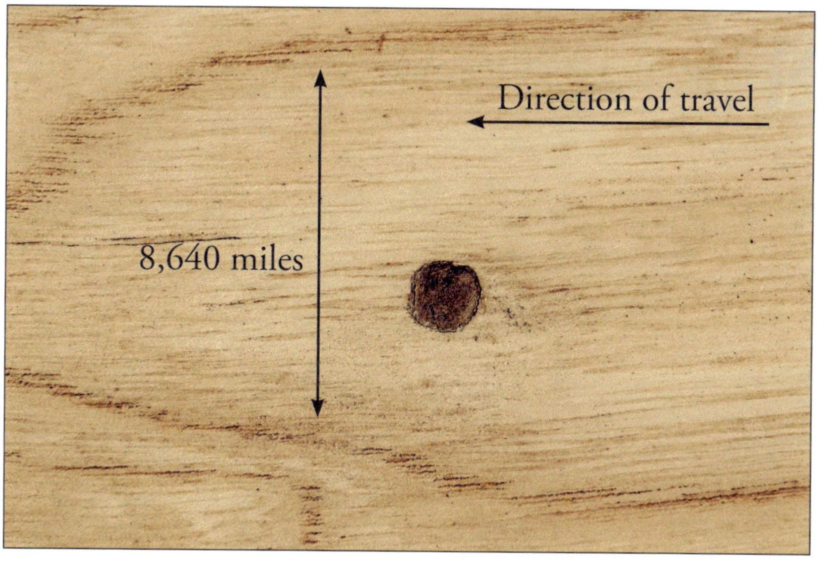

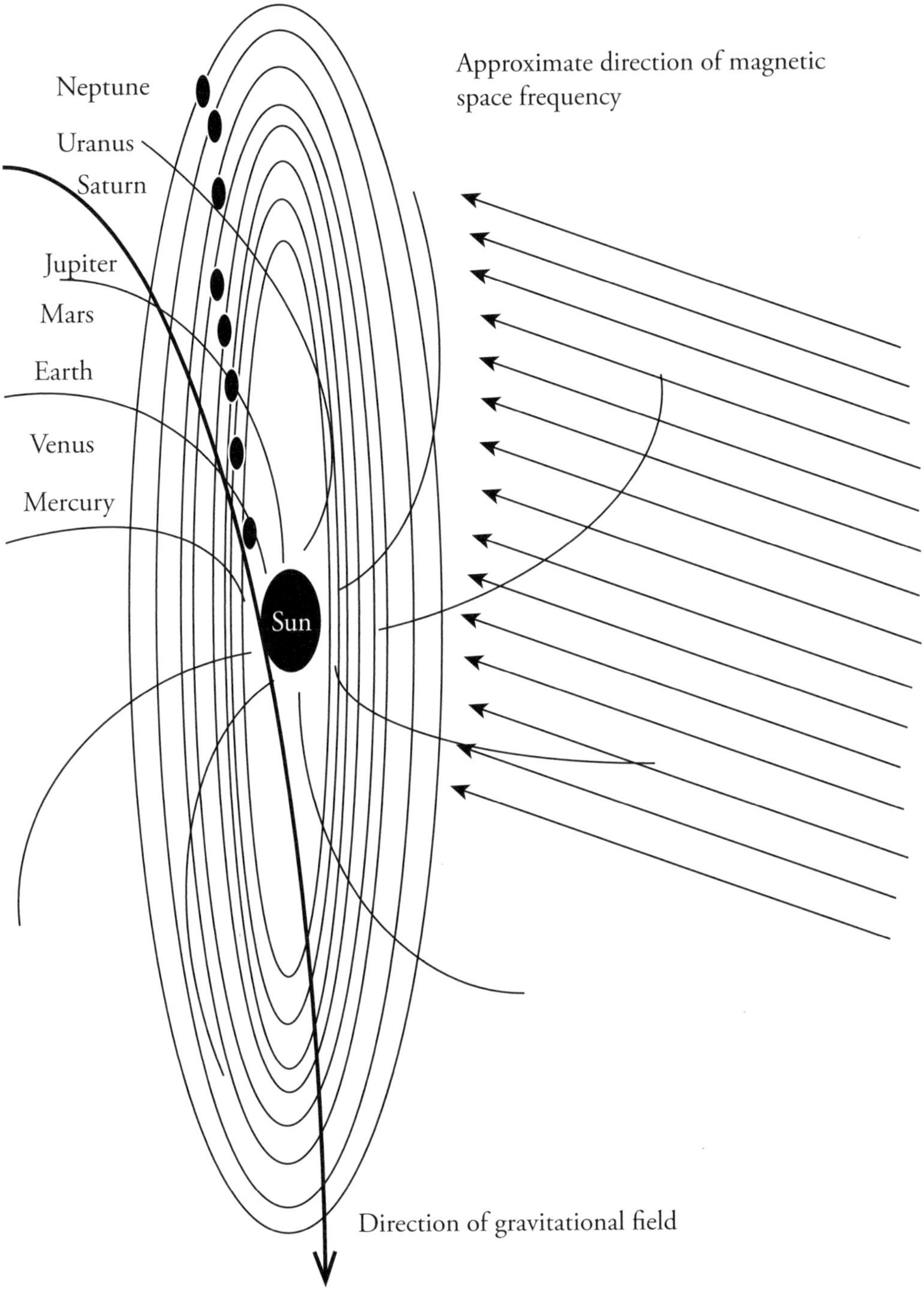

This diagram reflects direction and expansion of magnetic frequency plus the direction of the solar system.

Being raised in the country, I spent plenty of time climbing and studying trees. Later on in my life when I became interested in the tuning concept, I started sketching visions of planetary motion. In 2018, I was sketching the solar system and what I believe the magnetic field would look like being spun off of the sun's rotation (see page 43-44). The first sketch I drew was spinning counterclockwise looking down from the north. It wasn't unitl early 2019 that I noticed fence boards, plywood and veneer all had similar traits. I started researching my notes and found a sketch. I held it up to the knots and there was a match in frequency energy surrounding the knots representing the earth.

This became an interesting subject when friends came over to talk about my book. I would gather some specimens of wood that were photographed for this book and explain "now if we look at this knot, it represents earth spinning counterclockwise" as I point at the veneer. "These lines represent the magnetic field stacking up behind earth's travel around the sun." My friends first reaction is a type of blank look on their faces as their approach of understanding my concept became argumentative. The other things I noticed was their attitude toward me as being touched or a bit crazy. Nevertheless, I'm going to continue explaining these wonderful traits of all the common natural wonders of wood and associated phenomena.

One of the questions I've been thinking about since childhood is, "which came first, the chicken or the egg?" On page 43, the answer to me appears to be the shape of the magnetic field around the earth. The answer being the egg came first.

The above knot is the untouched photo which is approximately in the 8% category of classification. Plasma ejection.

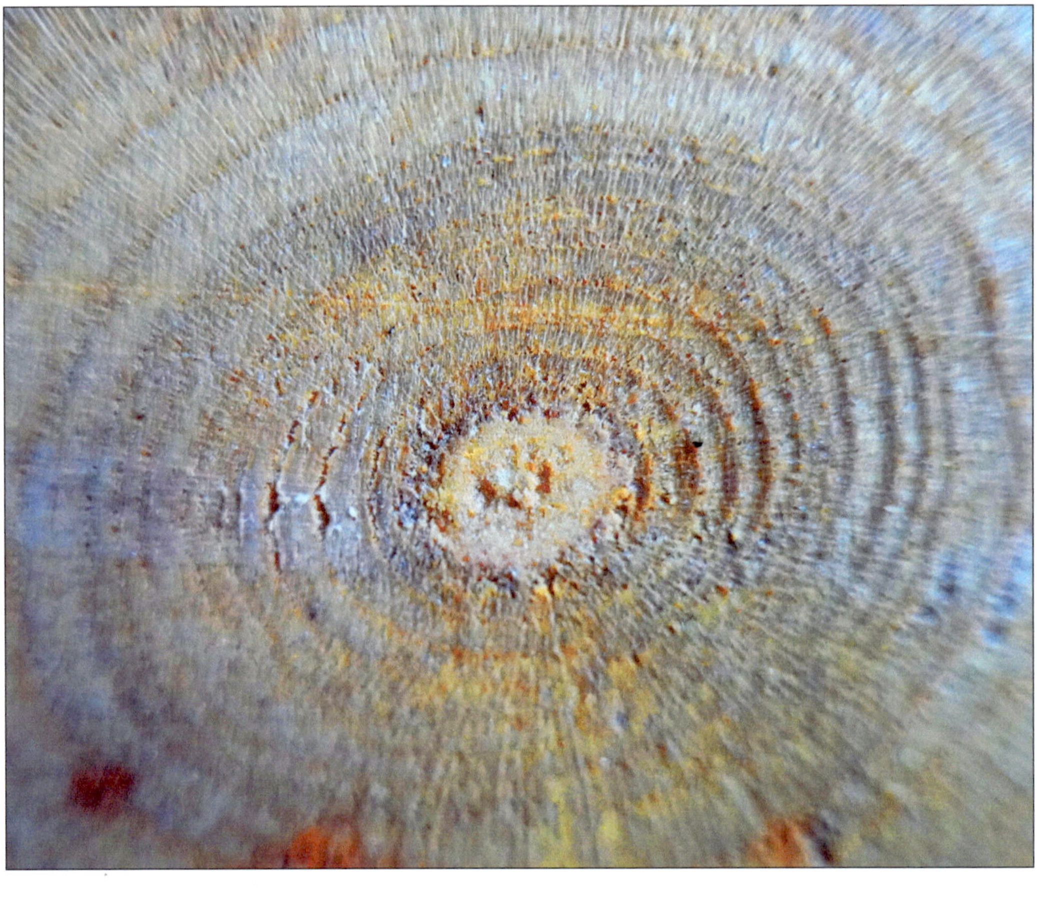

These three photos of the same knot seems to show the sun expelling plasma energy with energy rings being growth rings in the branch. The two colorized versions of the knot hint of different energy output. The blue line that runs horizontally is a curiosity as all of these photos are. The distance of the energy rings are approximately 125,000 miles apart using the sun as a measuring scale, then expanding. The biggest curiosity with this photo is the representation of plasma energy expansion.

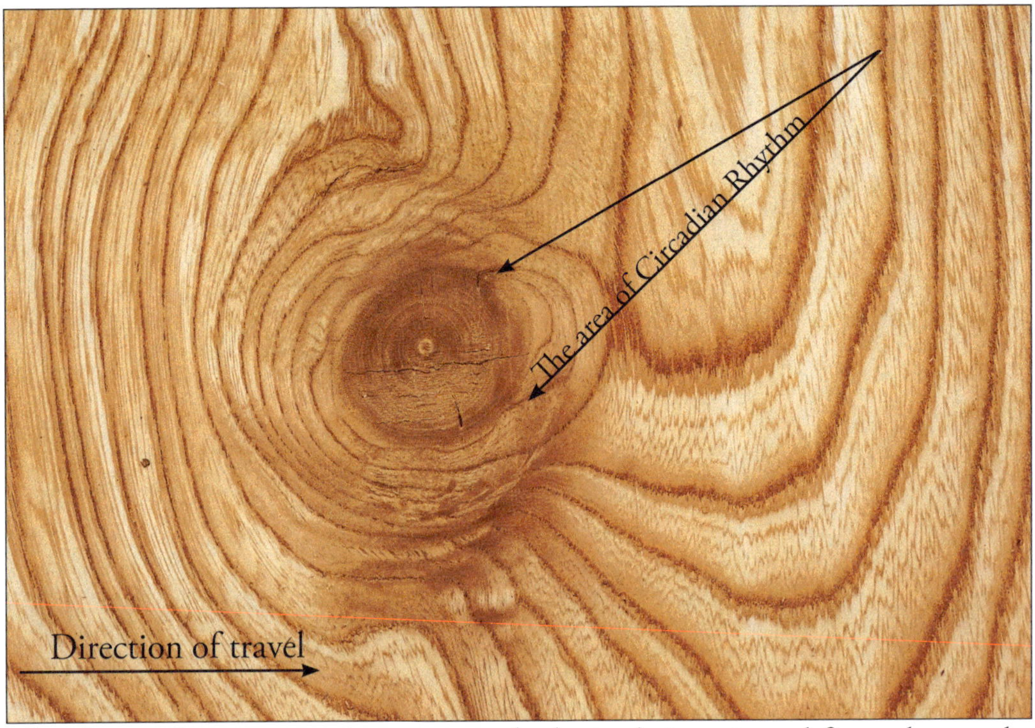

The above photo is spinning counterclockwise being viewed from the north.

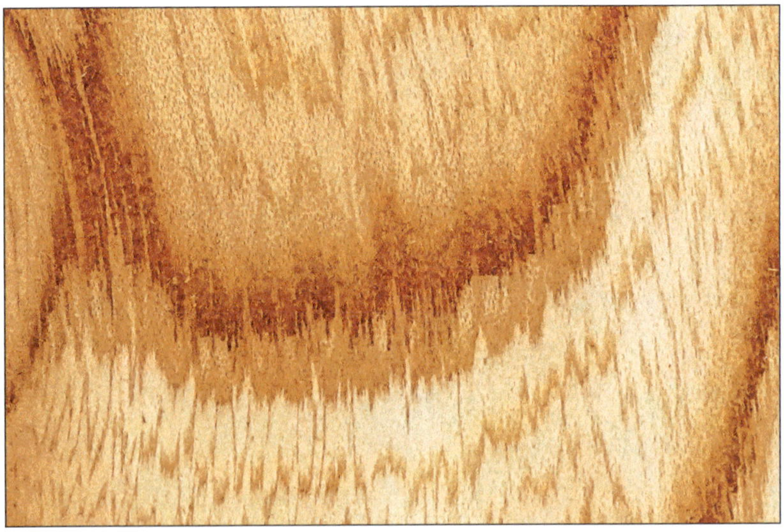

This enlarged photo of the top image is unique as it shows two different energies. Both energies from the sun represent frequency vibration. These energies are transferred to earth's magnetic field as shown in the top photo. My belief is the energy between the dark lines is low frequency and the dark lines themselves represent high frequency. (see page 15)

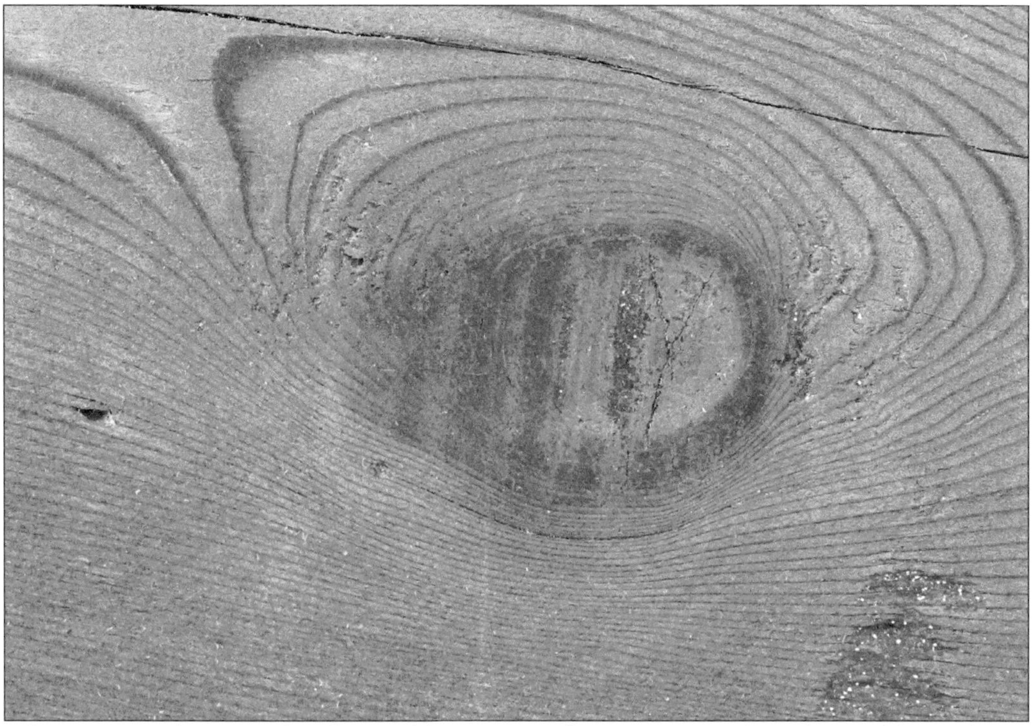

These two photos of redwood knots are from a 1"x12" redwood board dating back to approximately the 14th century. The tight grain of frequency lines represents a much colder environment being a measurement of the sun's energy.

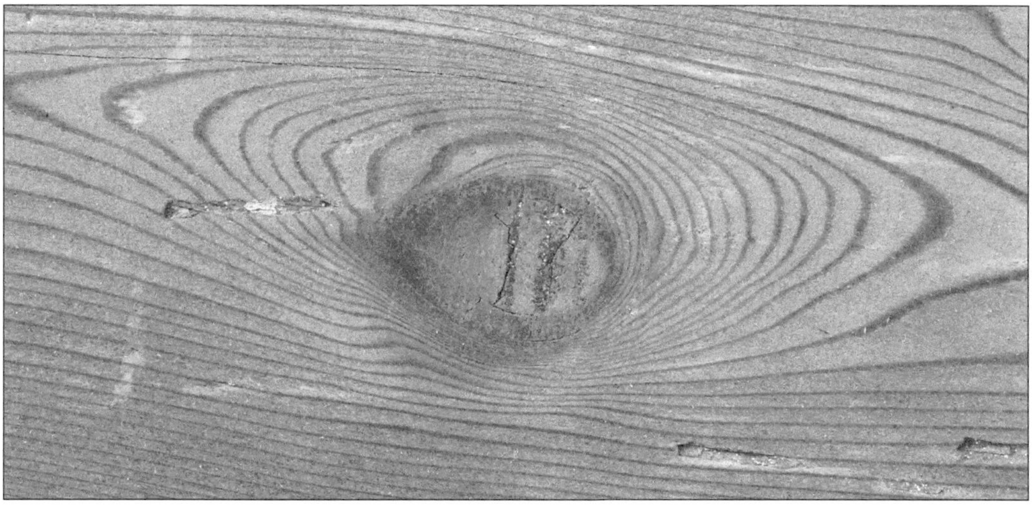

On the same board, the frequency lines show a slightly warmer environment. Both of these knots are spinning counterclockwise viewed from the north.

These next two redwood knots display energy distorting space as it suspends itself in a radiating electrical energy field. Note the differential between the two poles as the lower energy field is reaching out as a sensor-type of a looping effect. This photo gives the appearance of the energy field coming off the page in a three-dimensional distortion. These knots on pages 56-57 are in the category of the 8% and may be representing the sun.

This photo below has tighter frequency field lines on the top and sides reflecting a cooler pattern of energy.

The bottom of the photo produces a splashing effect of energy as it distorts space in a 3-D visual illusion as coming at you. This visual effect reminds me of a spider web that just happens to be to my left outside the studio.

The lower portion of the picture portrays an energy field that is being controlled from the lower section of this object, possibly representing the sun. I have to remember that this knot is from a 1"x12" piece of redwood and not veneer plywood. The difference between this sample of wood and veneer is the peeling effect of processing veneer.

The oblong distortion of this knot reminds me of an angle knot tilting slightly in the frequency of an energy field or possibly an eclipse. This classification is open to interpretation.

From the same 1"x12" piece of redwood, I call this the angle knot, upper left. I find this interesting as the smaller knot doesn't interact as most of the other knots do. If you follow the smaller knots frequency lines, they veer away and don't join the angle knots energy field. Another feature of the angle knot is the effect of it absorbing energy similar to a black hole. It may be that these diagnol knots in space frequency may cause a hole in the energy field. It's as if the poles lie vertical in space frequency or have two negative or two positive poles.

Interpretation of the knots like these is a study in imagination.

The picture above is a computer modification of contrast and color manipulation of the photo on the previous page. The reason I added this photo is to suggest that our eyes are only capable of seeing a very small portion of what is happening out in space. When we go back to the original photo, I see energy field lines, objects and relationships that seem to be from our solar system and deep space. When I look at the stars at night, I know that the sky is filled with energy.

These four knots have an interesting relationship in magnetic frequency. The two larger knots represent planets (Earth) and the two smaller knots represent moons. The moons have magnetic frequency moving around their mass. The planets have the same characteristics of the other planetary knots seen earlier in the book. I would think that the magnetic frequency would be able to penetrate the moons mass but instead it simply moves around its perimeter. The moon knot on the right appears to be moving right to left as the moon knot on the left seems to be moving left to right. The large earth-like knot on the bottom of the image appears to be spinning counterclockwise and the circadian rhythm is on the righthand side of the knot spreading through the decompressed zone.

Using logic and analyzing this picture below, I see this redwood knot moving left to right but also moving diagonally out of the picture toward my right. The tight frequency lines around this knot seem to suggest a slow movement through energy that reminds me of a slightly distorted moon.

Another look at this photo reminds me of a lunar eclipse.

Dudley of Dickson Ranch (Photo by David Russ)

I want to include the study of tree knots and horses as a comparison since they both exist in my environment. The photo on the upper right shows a knot which is spinning clockwise putting the sun's energy approximately in the center of the horse's forehead level with the lower portions of the horse's eyes if the lines above the horse's eye are considered as similar to the knot. Every horse has a single star or a double star on their forehead varying in location.

The same could be said for the horse's right eye as the energy would be spinning counterclockwise located in the same position in the middle of the horse's forehead. When I'm thinking about this, there seems to be a split down the middle creating the clockwise, counterclockwise spin of energy that seems to exist in most living things.

When I went for a walk to find a subject for this page, I looked at the horses expression on their faces and noticed this similarity. As I studied their expressions I instantly thought of a woman from India wearing a bindi, or a red dot, on their forehead representing the third eye.

Discovering a Change: Ratios

The reason for this new added section making the third edition to this book, is because the magnetic field has changed its two main ratios of its tuning system. Basically, I'm leaving the first part of this book alone because it explains the situation back around 2018 and some years before. I can now skip studying most movies and just focus on Martin Luther King's speeches of 8/28/63, *I Have a Dream* and *I've Been to the Mountain Top*.

The first thing I identified within the video of *I Have a Dream* is the song, *We Shall Overcome*, recorded, and over-dubbed for the peace march to the capitol on the day of his speech. This song remains in the key of D no matter, so far, what the magnetic field is doing. Many videos leave out the marching segment with the song *We Shall Overcome*. A/C = A over C

I mentioned notebook number 34 as studying the I Have a Dream speech as being in the key of A/C# (being 4 half steps apart was caused by the magnetic field changing its ratio) on pages 30 and 32. I was having problems identifying the pitch of the speech because the magnetic field was changing its frequency effecting Martin's pitch of his voice and also effecting the audience's clapping in response.

The first time I mentioned LR (long ratio), was on page 10. I wasn't aware of it but I was getting close to listening to soundtracks in their original keys when they were recorded.

I charted the speech on 7/26/18 and decided Martin's pitch was A#/C all the way through *I Have a Dream* speech. The A#

equals the low notes of his voice which puts his speech in A# and lines up on a lower octave frequency of the high magnetic frequency being five octaves higher or 5:1 ratio. The high notes of Martin's speech being C matches the low frequency of the magnetic field which is four octaves higher or 4:1. A#/C = C being the root is what key the magnetic field was in. Going up from A# being the root of his speech 2 half steps identifies the root of the magnetic field equaling C. This is different among my studies as 3 half steps down from the magnetic root then up one octave was the norm for the octave relationship of the high frequency of the magnetic field. I have notes on charting his speech and it falls in half setps during the earlier time period of study. This no longer exists.

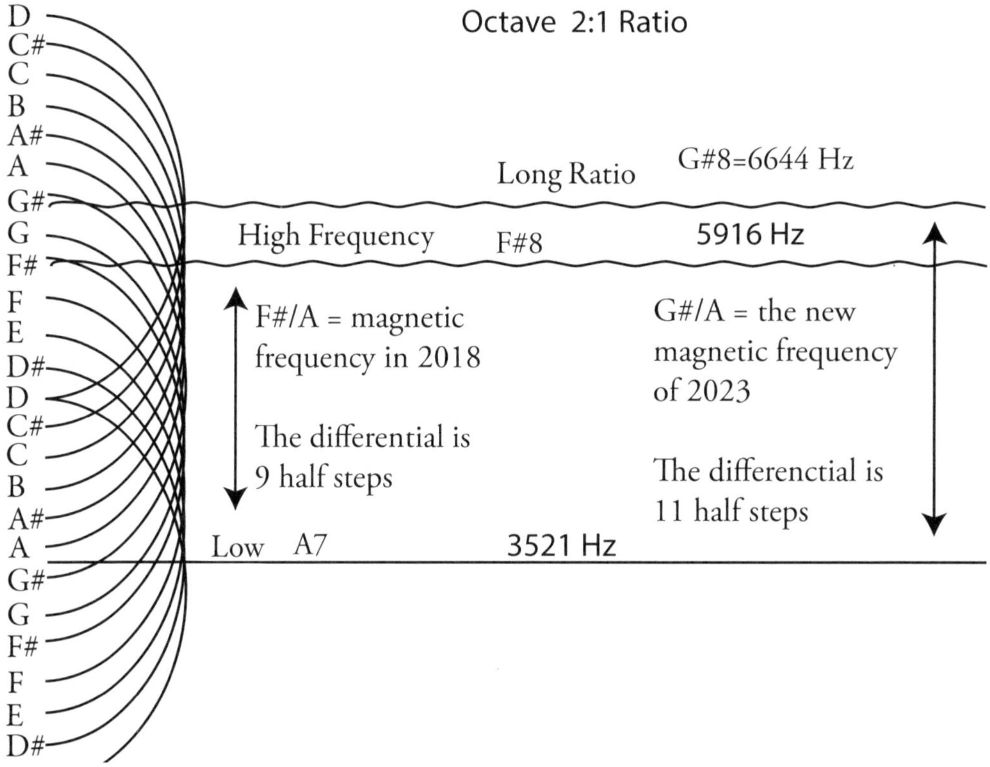

Charting the magnetic field's frequency at that time of 7/26/18 was F#/A then up ½ step to G/A# being three half steps down from the root of magnetic field of A or A# then up one octave to the lower divisible octave of the high frequency of the magnetic field, similar to the graphs on page 15 and page 64.

The closer the high and low are during his speech the longer the ratio is in the magnetic field which translates to speech (clocks ticking, doors shutting, people talking, etc.)

I have a chart on pages 21 and 22 of the mid frequency reacting to the low frequency which recently has not been as much as a factor as it was. So I'll concentrate on the two main frequencies.

The ratio of the two main energies is a glimpse into how a force field in the future could be converted or adjusted to accomplish different objectives such as repelling or absorbing other energies.

On 7/25/23, the frequency moved to G/G which changed to G#/G# both being 2:1. This phenomenon is recent, and I've

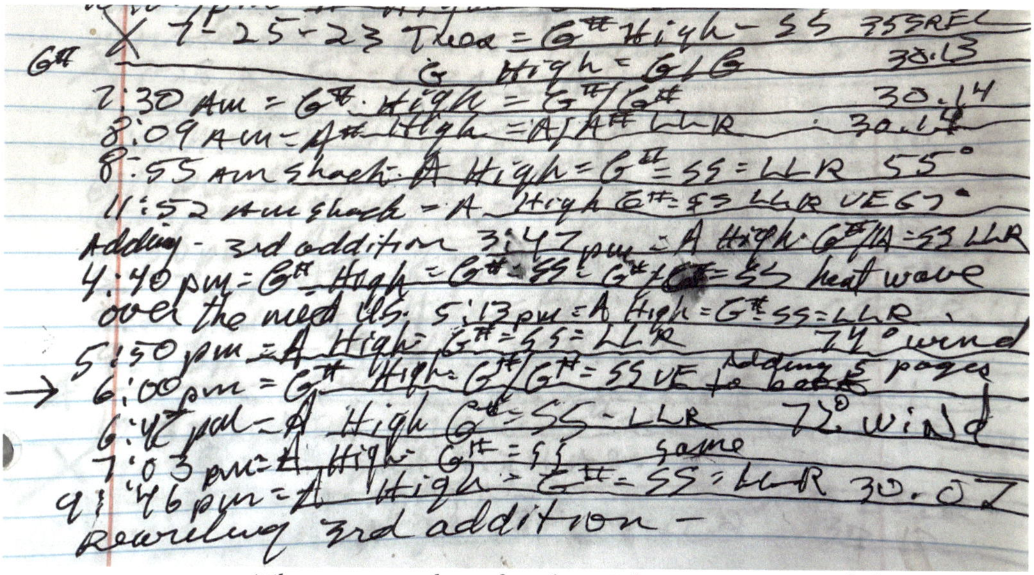

These notes identify when I became aware of the 2:1 ratio.

never charted this before. In recent events across the United States, there have been over 5000 new heat records set and scientists are noting the whole planet seems to be heating up.

Crow Talk - Dog Talk

When the frequency changed, the young crows were startled and asked the older crows to explain it. The older crows gave a relaxed lecture as if saying, "it's no big deal, nothing to worry about."

The G/G would be aligning the magnetic field to either allow more energy from the sun to come in or block the cold temperature of space to be absorbed into the atmosphere, or it could be cooling the atmosphere. I am still gathering evidence to draw a conclusion. As I am reading getting ready to work on the book, a neighbor's dog alerts me to this change of the magnetic field. 8/24/23 at 9:26 am the frequency changed to G#/G#. And then 5 minutes later the frequency went to G#/A.

I've been studying dogs since I was a child. When a dog notices a shift in the magnetic field they will analyze and express their knowledge by barking in code. On 8/25/23, one dog felt the magnetic field change abruptly from A/A# to G#/A and within a few seconds barked a code of G#/A = 2 barks a second on A then 1 bark on G#. The magnetic field had changed from A/A# rather abruptly. Again and again, the dog barked the same code – 2 barks a second on A then 1 bark on G#. This means the dominant frequency was the low magnetic note of A. The G# being almost an octave up from the low magnetic note was not as strong a frequency and only gets one bark. The dog identifies the differential exactly as I do. Being one half step apart symbolized the G# equaling the high frequency at 6,705 Hz and the low frequency of A at 3,730 Hz. This one dog in particular later barked a slightly different code, as the magnetic frequency changes periodically. My thought is over half of everything that's related through their barking has to do with the magnetic field and its movements, intensity, etc. Besides from using dogs for analysis, I've been using ravens and crickets to double-check my work.

Now in 2024, I'm reanalyzing movie soundtracks and finding the majority of them are falling into a new category of B/D/F/G# and this creates a diminished chord. The diminished scales are slightly different. Some of the movies may be a half step or a whole step sharp or flat of the central keys of B-D-F-G.

Ever since this ratio started to change around 6/29/21 in notebook number 40, I knew I would end up adding this new section to the book. This new magnetic frequency relationship now allows me to hear movies and speeches most likely in their

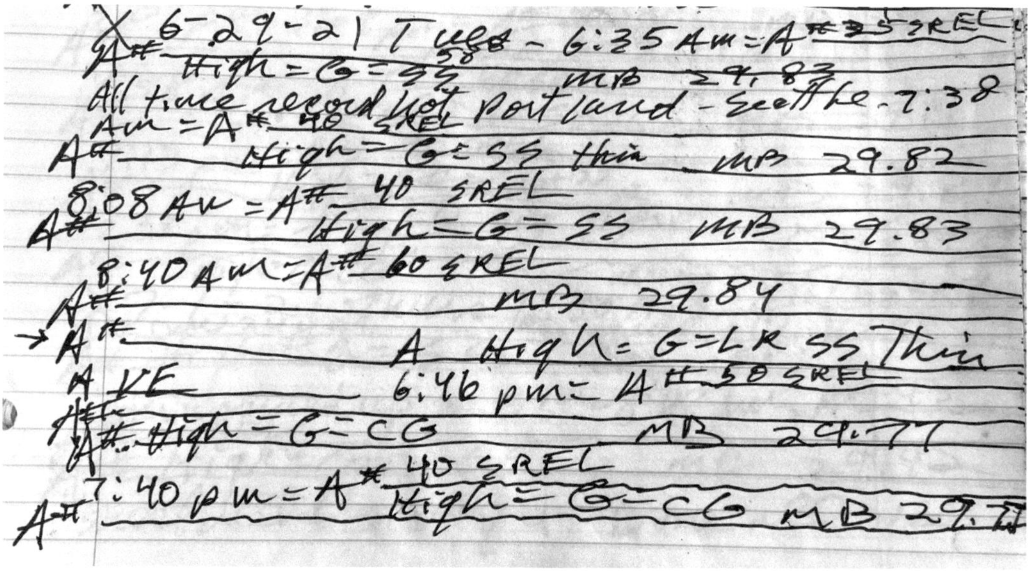

From Notebook #39, this graph is the beginning changes of a longer ratio from 3 half steps to 2 half steps differential.

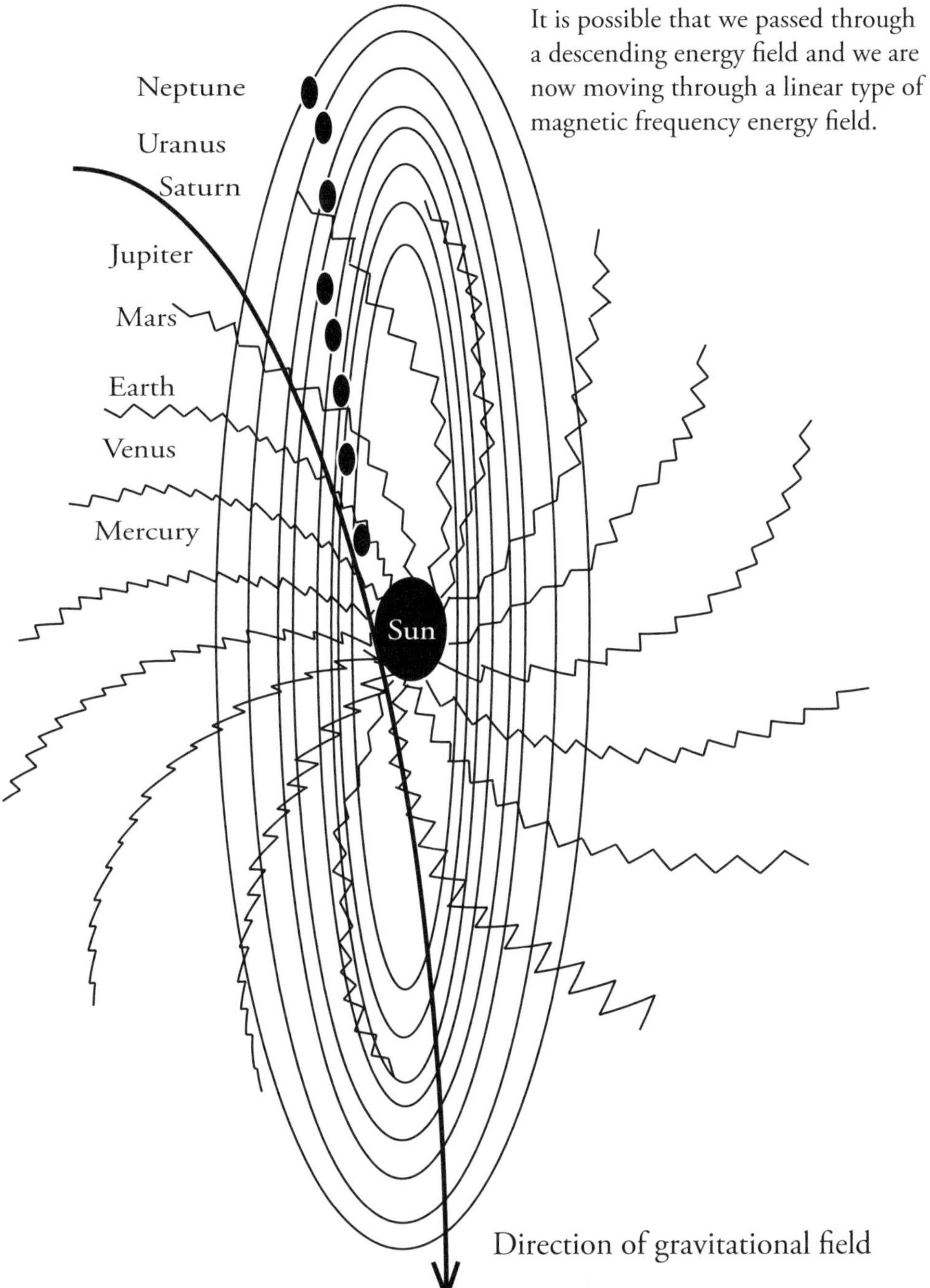

Each one of the magnetic frequency lines emanating away from the sun represents descending notes of the magnetic chromatic scale. As the energy passes the earth's magnetic field, the descending magnetic notes from the sun get transferred into our planetary system.

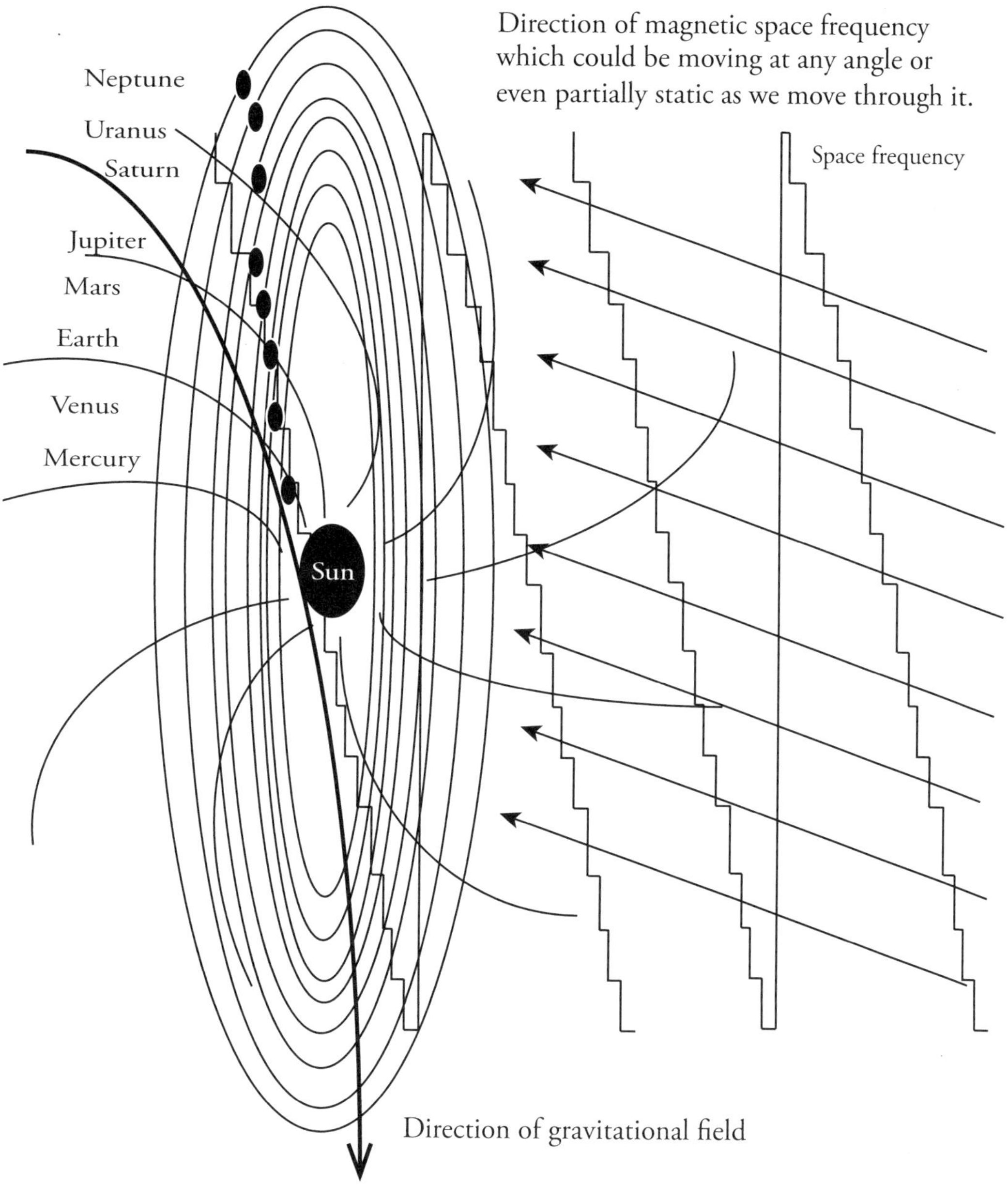

This graph represents space frequency. The question becomes - was it the space frequency that was effecting the sun to have descending magetic frequencies or was it space frequency that was effecting our planet to have descending frequency drops that I've been tracking in movies?

In 2018, when most movies were descending in pitch, this graph could explain the phenomena.

normal original keys when recorded or filmed. The song We Shall Overcome helps me to understand this relationship since songs remain in their original keys. This is why MLK's speech is important to me, plus realizing the plight of his cause.

Now when I chart *I Have a Dream* with my real time magnetic frequency being this new ratio, Martin is in the key of D/F and toward the end of his speech adds a high note of G# which helps create emotion and makes a diminished chord. There are many examples of the 3 half steps in this study. The number three exists everywhere when charting speeches and movie soundtracks.

A recent study of Martin's speech, *I Have Been to the Mountain Top* of 4/3/68 (this was his last speech), is mentioned on page 30. I again charted his speech on 6/5/18 and it was F#/A, E/G, D#/F#, D/F, C#/E descending all the way. I charted the same speech on 7/26/23 with the magnetic frequency at the time being G#/G#. The result was his voice stays on F/G# all the way through the 43 minutes of his speech equaling the 3 half step differentials.

This new differential of the magnetic field creates new harmonies as the original frequency I was working with in 2018 equaled F#/A which creates a 6th in musical terms. G/A being a longer ratio creates a dominant 7th. G#/A creates a major 7th. G#/G# creates an octave or 2:1. So far, the magnetic ratio of G/A - G#/A – G#/G# or any combination that have two half steps or less allows me to hear recordings in their original keys.

Early Grade School

Most musicians use the major scale to explain the layout of songs and chords. Everything fits together on a logical equation. Simply put, the major scale, do re me fa so la ti do (learned in early grade school for most of us) is the main formula. Each note is assigned a number. Do equals 1, Re equals 2, Me equals 3, Fa equals 4, So equals 5, La equals 6, Ti equals 7 and Do equals 8 which equals one octave or 2:1. The ratio between semitones equals 1: (twelfth root of two). C4 equals middle C on the piano = 261.63 Hz x 2 = 523.2 Hz = C5 on the piano one octave higher. It's interesting to think we are listening to various vibrations of mathematics.

The basis of the magnetic field and music revolves around the 12 notes of the chromatic scale. Flats and sharps are inserted in the major scale to include the remaining 4 notes which then equals 12 notes. When counting up 12 notes, it stops short of the next octave by 1/2 step. The last 1/2 step then makes the octave or 2:1. Each key has different flats and sharps. In my case, I'm using a tuner that works off the key of C which has no sharps or flats in the major scale. I also used a tone generator from a computer to help locate what frequency I was sensing.

It's possible the shape of the magnetic field may be altered when changing its frequencies. I'm beginning to run experiments on this subject. Going to page 41 explains a theory in the examination of the tree knots and the possible relation to the magnetic fields patterns of energy.

Overview

This new information points to two overviews of thought – one thought is as the magnetic force field changes its dimensional energy ratio – this may allow more energy from the sun to penetrate the atmosphere, raising temperatures. As the magnetic field is a force field, this may show how two energies can change the reflecting or absorbing factors of the field. The magnetic field may be reducing the temperature of space, being -454 degrees Fahrenheit or -270 degrees Celsius from being absorbed into the atmosphere. If the magnetic field works one way for heating most likely it can work the other way for cooling.

The second line of thought is the sun's energy may be mea-

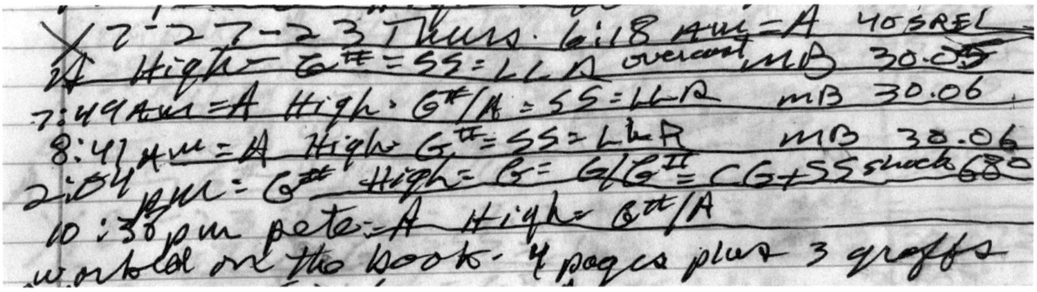
12-27-23 Thurs. 6:18 AM = A 40 SREL
A High = G# = SS = LLA overcast MB 30.05
7:49 AM = A High: G#/A = SS = LLR MB 30.06
8:41 AM = A High G# = SS = LLR MB 30.06
2:04 PM = G# High = G = G/G# = CG + SS shock 680
10:30 pm pete = A High = G#/A
worked on the book. 4 pages plus 3 graffs

sured as energy through the magnetic field as it affects most everything that happens to the planets. When the magnetic frequency drops in pitch, this shows more energy is produced if it is similar to a bass or guitar amplifier measuring energy output as in pitch frequency. Also the high frequency expanding its ratio to G/G adds a new factor to this study. The G/G is not engaged full time because most of the time the new magnetic frequency keeps a differential between the high and the low of ½ step or a whole step differential.

Somehow, I'm still convinced the Mayan calendar plays a part in all of this.

I checked the honeybees on 7/27/23 and they were all on G - G# as the field changes ratios periodically.

I charted some old Gunsmoke TV shows from the 1950s and found them to be in the key of D/F. In all my older movie books, these shows descended in pitch. I watched a Gunsmoke show dated 4/8/67, called Nitro Part 1, and it was in the key of D/F. All these movies have a differential of three half steps down from the magnetic root for the vocal tracks and one

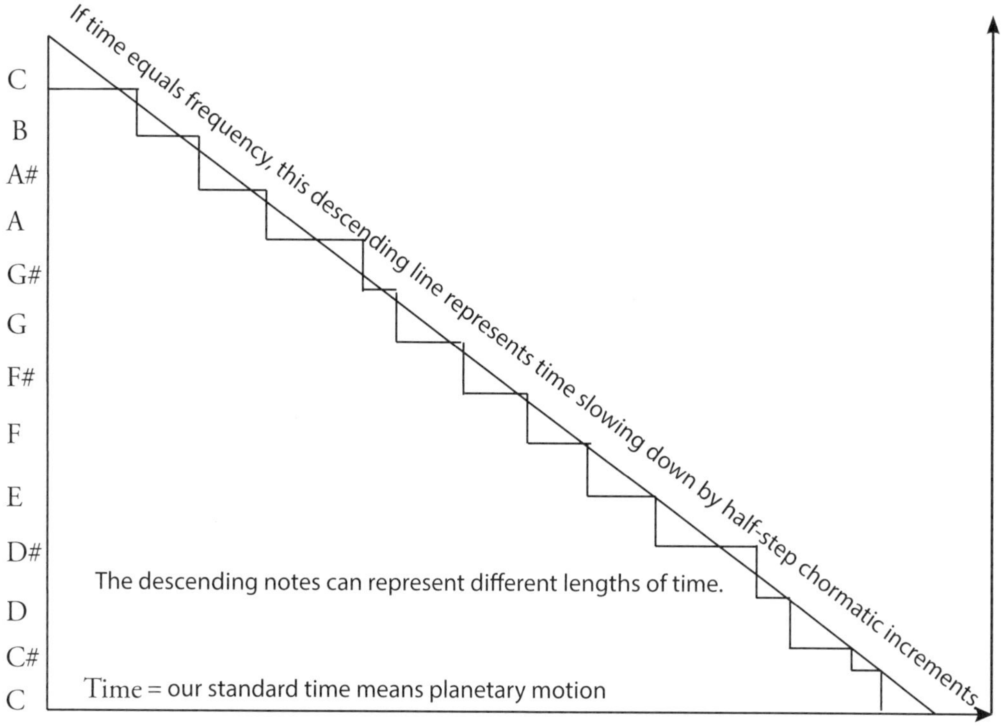

We took this graph from page 31 to represent my feelings of when movies were descending in pitch frequency. It always seemed to me that time was slowing down in half step increments as pitch dropped.

octave up locating a similar octave for the high frequency of the magnetic field. Another Gunsmoke from 12/20/58 called Snake Bite equals E/G which is one whole step higher than most average movies. The music in the show lines up perfectly with the vocal track. At that time on 8/9/23 when I charted this show, my magnetic frequency was G#/A which equals 1 half step differential.

It's becoming obvious that the chromatic scale of the magnetic field causes pre-recorded speech and real-time speech to fall into categories based on mathematic ratios.

By not having a conversation of the magnetic frequency and its effects on global warming and cooling is very short sighted. But here I go again using sight instead of hearing and sensing.

The new changing differentials means peoples vocal response and other entities keeps the same pattern as the real time magnetic frequency field of energy.

If the pitch frequency of the magnetic field drops and the energy remains the same in my ears, does this mean more energy is being produced? I think, yes.

As my mind examines the probabilities of discovery, there is another thought which in the ratio factors allows me to speculate that even though the ratio changed, the earth could have passed through a force field of descending pitches.

If we passed through the descending space frequency and now we're in a modified linear pattern of energy, this may have caused the magnetic ratio to change.

A note: I've always been thinking of a way to deal with the extreme fire hazards. A futuristic thought may be to bombard the fire area with electronic frequencies that are opposite of earth's magnetic field to confuse the chemical combustion.

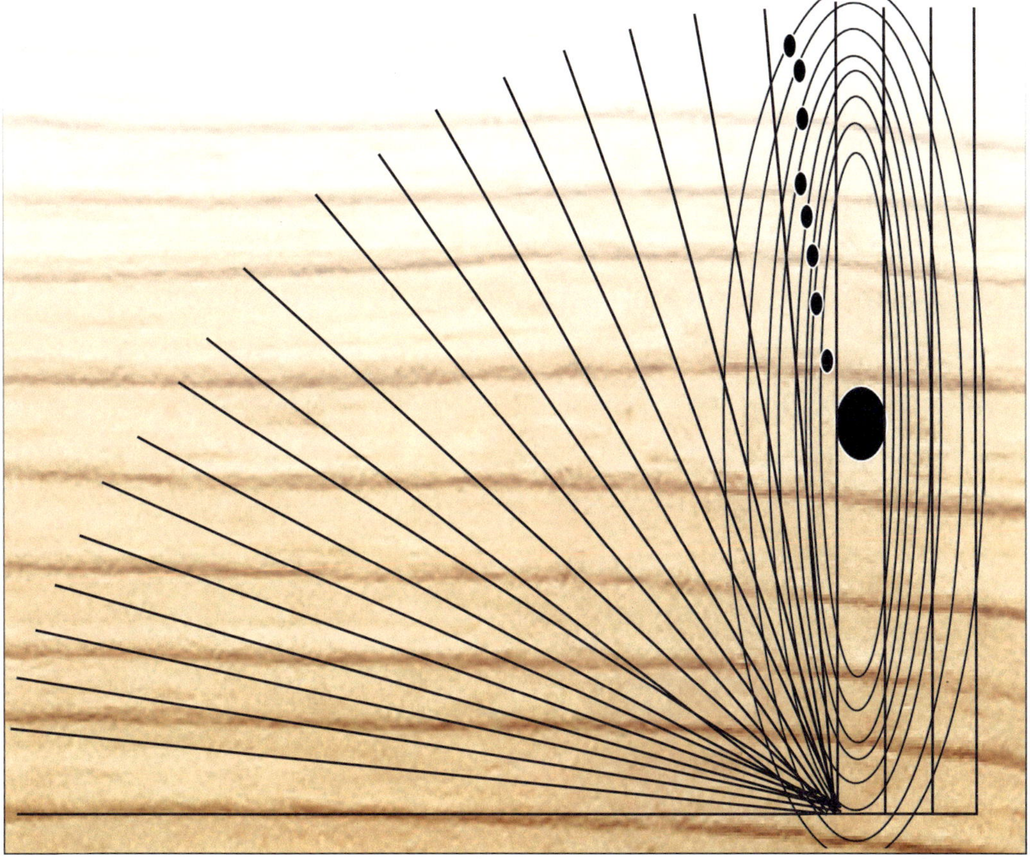

This wood grain represents space magnetic frequency as it relates to the solar system (enlarged).

There's a possibility as we cross the energy field lines it causes movie soundtracks to descend in pitch frequency. I was looking for an explanation of why we are in a more or less static relationship that holds most movie soundtracks in the key of D/F/G# which is a diminished chord. Our trajectory now appears to be that we are traveling in between the energy fields creating this diminished frequency of 3 half steps apart. The graph suggests a point where the energy lines converge and may be hundreds of light years away.

This brings us back to the idea of the Mayan long count calendar. Their idea of a change could be related to this concept.

If our systems trajectory was at a right angle to space frequency, the magnetic frequency could drop in pitch as the graph on page 34 suggests. The dark lines of energy could be a reset frequency jump to the octave to start over from the high to the low in half step increments.

On page 42, on #3 of the picture, the lines of energy changed and are running basically 80-90 degrees from the other energy lines which if our trajectory was going in the direction of #3's arrow would be basically static frequency. The frequency lines between the dark energy lines represent notes of energy.

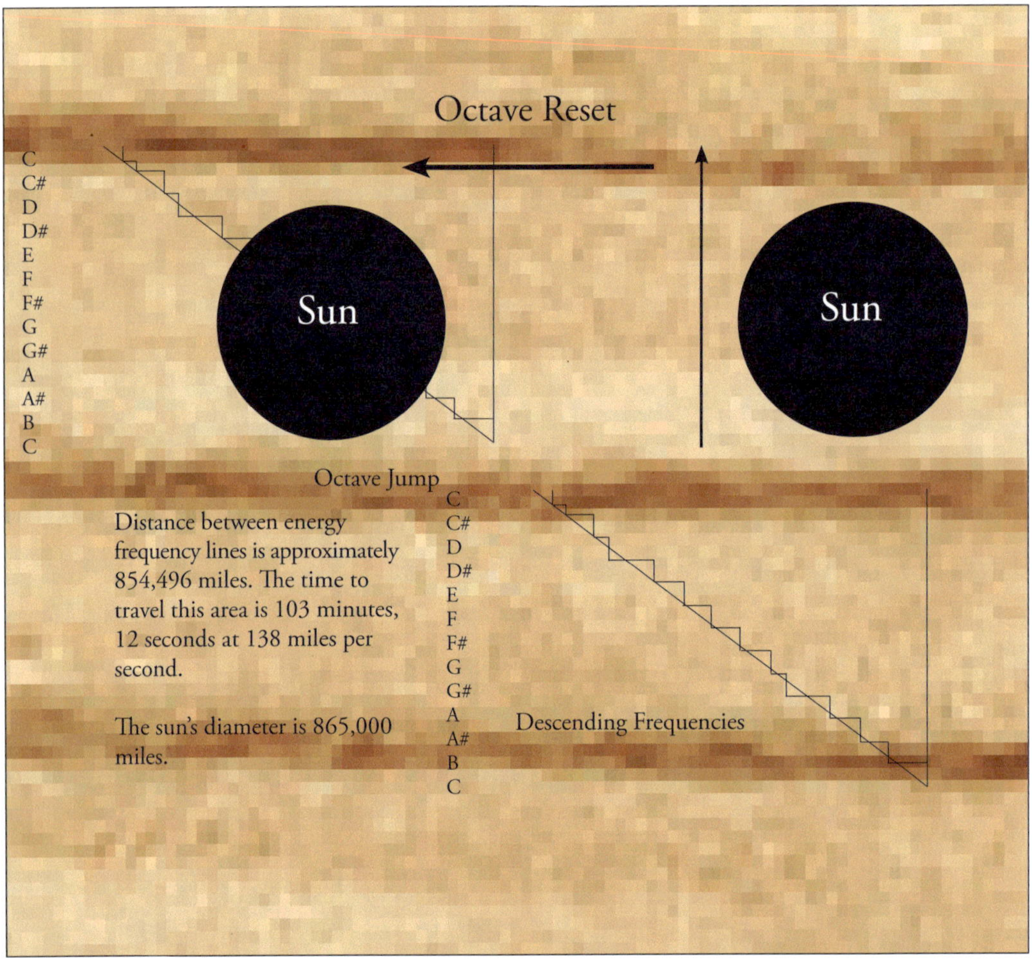

As the sun engages the chromatic half step magnetic field energy, it breaks the energy down into three half step diminished scales equalling one octave.

Calculating

I wanted to figure out the distance between the magnetic frequency lines on page 76. Also I wanted the speed of our solar system as it traveled through space frequency. This might explain if magnetic space frequency is static or moving.

To start calculating, I took Martin's speech, I've Been to the Mountain Top. The speech had five half step drops of pitch in 43 minutes back in 2018. I knew the magnetic field can travel a whole octave of frequency drops because I charted them around 3/11/2011. This is when Japan experienced an earthquake of 9.0 on the Richter scale. I divided 43 minutes by 5 half step drops equaling 8.6 minutes which equals 516 seconds per half step drop.

I then multiplied 516 seconds by 12 half steps equaling one octave of the chromatic scale which equals 103 minutes 12 seconds or 6,192 seconds. Then, I took 6,192 seconds times 138 miles per second equals 854,496 miles between the magnetic energy lines on page 76 (The 138 miles per second was calculated from page 23 in the book "Energy Changing Frequencies").

We have distance 854,496 miles between frequency lines – we have speed at 138 miles per second or 496,800 miles per hour – which matches the estimates of our solar system traveling through space. Estimates vary on the speed of the solar system from 514,000 miles an hour to 448,000 miles per hour. We have the sun's diameter of 865,000 miles putting the sun approximately inside the space frequency energy lines.

The energy I was able to time between New York to California was always within a 1.5 second variation. When spending hours upon hours listening with a tuner in my hand in front of a television waiting for the exact moment a commentator would drop a half step down in pitch while talking took me months. I had to wait until that very second to start a count on when I could sense the energy's arrival and catch it with my tuner. It was exciting and I had the experience many times but only felt I really got it exactly a few times on the mark.

I have the estimated time from 2009 when movie tracks went from dropping in half steps to going to the key of a basic parameter of G# A A#. This is the beginning of the augmented mode of the frequency field. In 2023 all recorded frequency (except for music) goes back to the original pitch of the recordings. The music tracks line up with the vocal notes of pitch. Music tracks don't change from the magnetic field so I can use music tracks to double-check vocal tracks.

From 2009-2019, is the transitional time between the change of half step drops to the augmented mode of three half steps. It took approximately ten years of time for the transition to happen. Ten years of traveling at 138 miles per second equals 315,360,000 seconds x 138 miles per second of travel equals 43,519,680,000 miles of travel through the change.

The difference I mentioned earlier in the time period of pre-2009 to post-2009 and earlier is now completely different as all vocal tracks are more or less in an augmented mode. All movie vocal tracks line up with the music tracks as before 2009 the vocal tracks would drop by half steps away from the music. Almost all of the music tracks in movies are overdubbed at a later date which is designed for the scene of sound.

The whole difference is the frequencies has now changed its ratios which start a new point and allows me to measure from the 2009 period to the approximate date of 2024 into the augmented period.

Waves at the Beach

I found the easiest way for me to find the relative acoustic notes that shows what notes the magnetic field is in. It's actually very simple. When relaxed I breathe in and find that note on my tuner. On 1/17/24 at 7:19 pm, I inhaled and the note was G#. This note of G# represents the low note of the magnetic field. Next, I exhaled and found that note on my tuner and that also is G#. This is relatively rare to have the notes match but recently it's more of a common occurence. This is the 2 to 1 ratio I mentioned earlier as G# over G# or G#/G#. Usually the exhaled note is lower by one half step or two half steps. The exhaled note equals the relative octave of the high magnetic frequency note. This simple rule can be useful in listening to many of natures sounds.

When I visit the Pacific Ocean out west, I can sit by the breaking waves and match my acoustic notes of breathing to the sound of waves coming in and out. I can say that most likely on 1/17/24, the incoming waves break on the note of G# and the retreating waves thereafter are also on the note of G# with a more relaxed energy of being in retreat. Usually in the past, the breaking waves and the retreating waves would have at least one half to a step and a half differential between the two sounds of energy. Now the ratios are acoustically closer at the beach meaning the opposite for the magnetic fields ratios.

Even though the graph of the sounds of the waves doesn't show in particular the high and low sounds I'm sensing as in A/A# from the magnetic field. This might mean my mind is focusing in on those particular notes because I'm being saturated by the magnetic field.

Birds

The pictures on the next two pages are taken from my computer at the studio. The computer has a music recording system that allows me to analyze soundtracks. The spectrum meter shows a picture of a recording of a flock of blackbirds singing on the ranch. Twice a year, the ranch is visited by several flocks of blackbirds as they rest for their next adventure. Through the years I can remember the black birds singing and realizing their songs are part of a great mystery. Little did I know that they would play a part in scientific history. If I listen closely, every bird and every animal tells a story about the magnetic field.

The graph shows two vertical lines which are on A#6 and A#7. A#6 equals 3718.7 Hz on the longer vertical line. The shorter vertical line towards the right of the graph is A#7 at 7437.4 being twice the Hz of A#6 or one octave higher. The day I recorded the birds the magnetic field was A/A# but changed slightly through

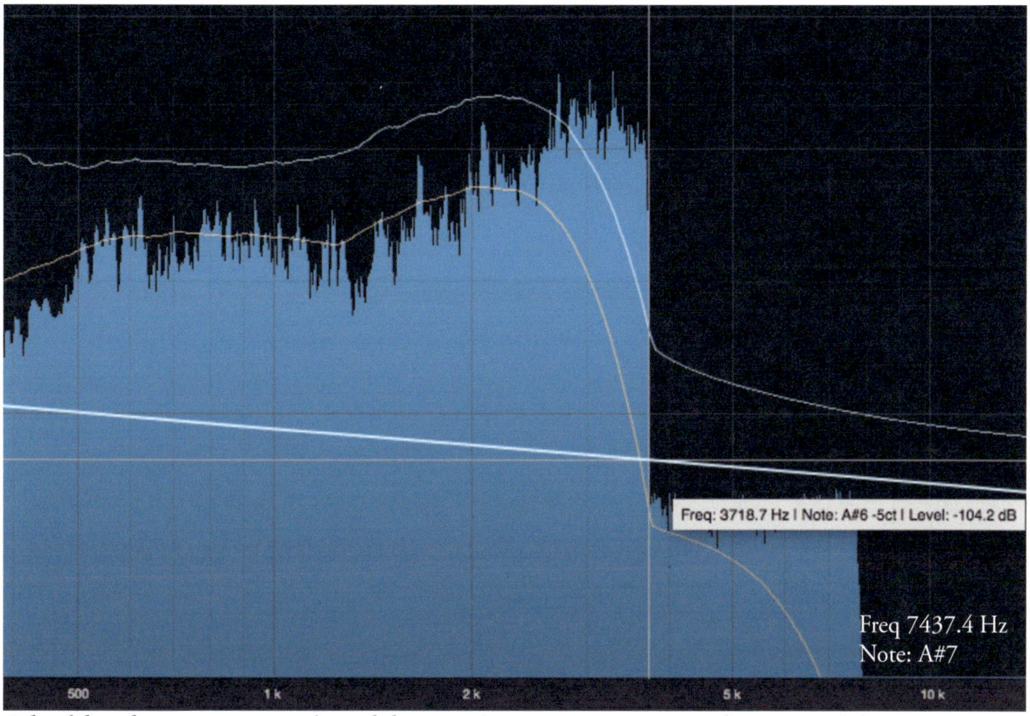

Blackbirds singing analyzed by a spectrum meter. Both vertical lines are A#6 (Freq 3718.7 Hz) and A#7 (Freq 7437.4 Hz). The double vertical lines could be associated with migration.

the day of Tuesday, January 30, 2024. My notebook shows that the magnetic frequency was A/A# in the morning and G#/A in the afternoon. This would put the birds in the magnetic frequency of A# which was in the morning. This equals the low frequency of the magnetic field at 3718.7 Hz.

I'm lucky this year of 2024 that the hard rain has kept the flocks here longer than normal. In March of 2024, we have 128% of normal rainfall out in West Marin.

The A#6 (A sharp 6) and A#7 stands for the octaves based on the piano starting on A0 (A zero) being the lowest A on the piano. The next A up above the first C note is considered A1.

People

On February 10, 2024, while at a friend's party, I noticed the chatter of all the people conversing which reminded me of something. It took about 5 minutes of listening to figure out it had something in common with the birds. I then started to record the conversations and with some help from friends transferred the recording to the studio computer.

The spectrum meter showed the results being similar to the black bird's graph but only having one vertical line at the frequency of A#6 being only 21.2 Hz differential between the black birds and the people talking.

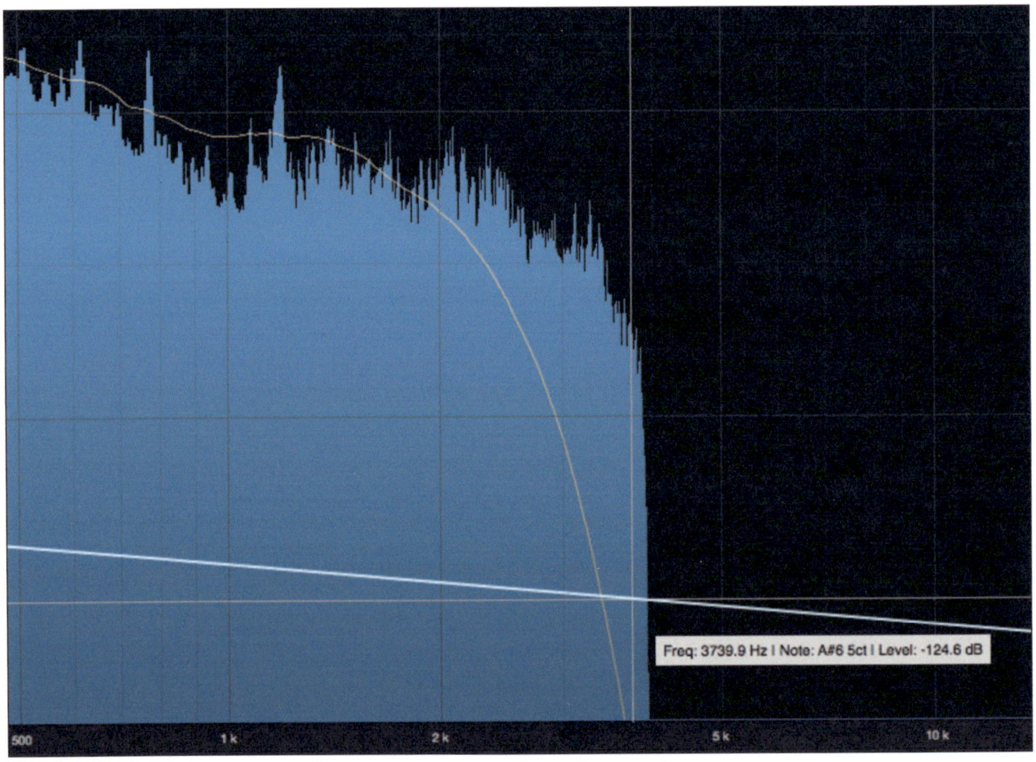

This is a graph of people conversing taken from the spectrum meter. Apparently, humans have one vertical line and are limited in the understanding of migration.

The graph shows the influence of the magnetic field throughout life on the planet. The birds have an extra octave of vocabulary compared to the people's graph. Both graphs are also analyzing language. To break language into logic, I have to establish common denominators to create analysis. I believe the graphs establish this. This is an analysis of nature and natural phenomena. This phenomena effects the microscopic and the understanding of how the universe operates.

Limantour Beach

The odds of the birds singing, people chattering at a party and the key of C being expressed in the the ocean waves is facinating. All of these energies are locked in to the magnetic field. The key of C may be pointing to a galactic energy which I'm in the process of understanding through scientific research. If it is in the ocean, most likely, it's in me. The challenge is to find it.

Limantour Beach, Marin County, looking Northwest towards Pt Reyes Peninsula. (Photo by Walt Dickson)

I've given a lot of thought to the saying, Space, the final frontier. But in the late 1980s and early 1990s, I started to believe that the final frontier was in our minds and bodies. If I can understand my body and other animals behaviors, I might be able to answer most questions associated with our galaxy.

If I'm from the stars then somewhere within me there could be answers of apparently unthinkable depts of understanding.

There was one thought that gave me the idea for this book. The idea was to view and feel planet Earth as if I was from another part of the galaxy. I saw myself entering the solar system and viewing earth for the first time and asking one basic question. Where does life get its information from? The tuning system turned out to be the most logical answer in its most basic form. The graph below shows C9 as being a major factor in the analysis of the waves. This C9 may be a sign of a magnetic galactic note.

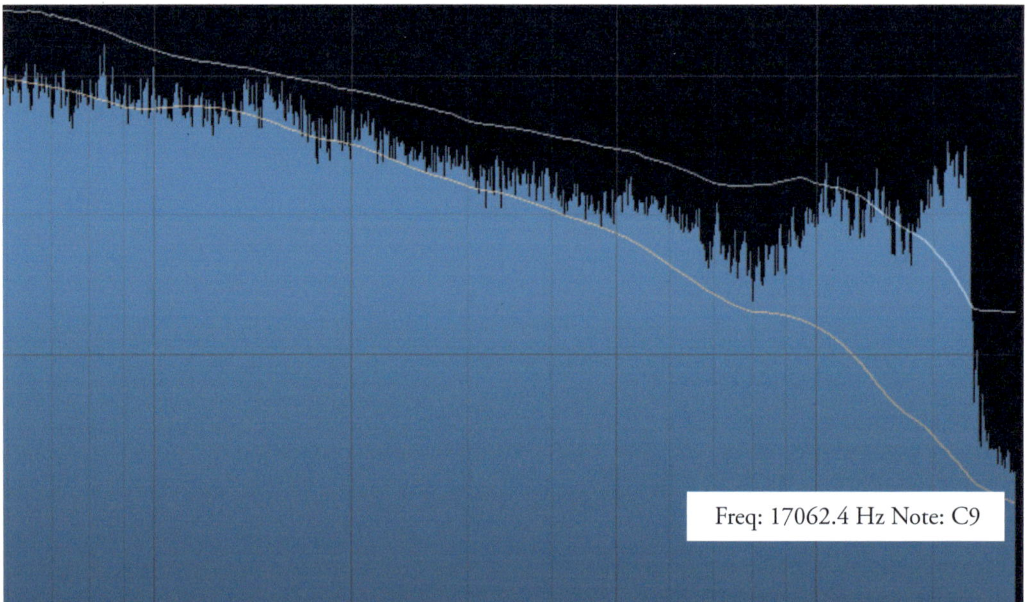

Here is a spectrum meter analysis of Limantour Beach in Marin County.

Influences

In my early days around 1960, the music scene was alive and growing in West Marin County. This was the end of the real cowboy days and the last of the living old west storytellers. The beatnik scene was happening and words like "cool" and "hip" were making an appearance. AM radio was playing Elvis Presley, Chuck Berry, Ray Charles and Johnny Mathis with rock 'n roll gaining momentum. There had always been a classical and opera scene from San Francisco. I remember as a child being at the club Bocci Ball in San Francisco where opera singers and musicians would carouse.

One of the first interesting musical people who moved to my family's ranch was Fae McNally, a classical pianist, who

l-r Greg Longfellow, Fae McNally, Thelma Dickson at the Inverness Music Festival, Inverness, 1970 (Rud. Ibach Sohn) ended up being music director of A.C.T. in San Francisco. Fae had met another piano player, Ralph Sutton, who played a small gig at a local restaurant. Ralph was a solid stride pianist and Fae soon started playing old jazz classics. Around this time, Fae started the Inverness Music Festival which gave Sky Blue Band a chance at playing standards with some of the top local players. Mickey Hart and Dan Healy started recording some of the shows which I've never heard.

Fae would feature a classic, like *Madame Butterfly* which my mother sang, to having Norma Teagarden, Harry Alto, Turk Murphy, Earl Hines, Ralph Sutton, Barbara Sutton Curtis and many others featured in jazz concerts.

Gene Daly

At about the same time, a carpenter, Gene Daly, moved onto the ranch to help out on building projects. (I believe his daughter, Adair Lara, has written extensively on his habits). He would buy Ford and Oldsmobile station wagons, cut off the top and build beautiful campers on the back where he lived. Gene loved playing guitar and practiced A.C. Jobin and other standards. When he was in the mood, he'd pull his chord book out and show me fingerings and techniques. He was a master carpenter and it was Gene Daly who pulled a piece of eucalyptus firewood from my brother's firewood stash and rebuilt Jerry Garcia's guitar. After he was done, the guitar is probably structurally stronger (see photos on pages 94-95).

Forest Knolls

In 1967, my brothers started a garage business in Forest Knolls where I worked pumping gas for a year. The rock music scene was exploding as people like Janis Joplin and Big Brother, the Grateful Dead, Ramblin' Jack, Mike Bloomfield, to mention a few, moved nearby and ended up buying gas and having their vehicles serviced at the garage. At the same time, we had our still unnamed band set up there so after hours we could practice. With an upright piano on site, Mike Bloomfield loved to stop by and play his favorite songs. Mike was one of the top guitar players at the time. I was impressed. I was learning to watch and listen. He was always in a rush but took time to put new strings on my guitar.

l-r Ned Appleby, Jeb Harrison, Nick Sutton, Walter Dickson, Corte Madera 4th of July Parade, 1970s

Elvin Bishop

Elvin Bishop moved here about 1975 and we're still joking around with music and working in his wonderful garden. He's been a thoughtful friend with a sense of humor and has a heart for the blues. (It was through Elvin that I met B.B. King). I can recall many times with Luther Tucker, Elvin and myself playing guitar all night long. I knew it was a long night when I walked out of the studio and the birds were chirping.

l-r Walter Dickson, Elvin Bishop, Barry Lowenthal, Terry Haggerty, David Russ, Dickson Ranch, 2018 (Photo by Miriah Fox)

Specks Irish Coffee

Around 1975, the band started playing Friday nights at Specks, a bar in Lagunitas. This became a hang-out for all kinds of musicians from Eddie Adams to Charley Spivac. Although the bar was small, we could have ten players jamming. We played jazz standards and blues with a developing horn section led by Ned Appleby (Beefy Red and Dicks at the Beach).

Phil Wood and Randy Smith (Mesa Boogie) started playing in the horn section and soon the standards included *Nica's Dream* and *All the Things You Are*. Phil being a music teacher and Nick Sutton playing piano gave the stride feel of rhythm with modern chord progressions. The other stride players were Anthony Wright, and Dan Dickson playing Jelly Roll Morten to Fats Waller and Bix Beiderbecke.

Phil Wood, trumpet & keyboard (Photo by Frank Lewis)

Luther Tucker

This time period opened another door where a guitar player named Luther Tucker appeared on the scene. Right away we became friends and soon we were doing shows everywhere the band played.

His guitar playing was original with flurries of riffs. Luther was well-dressed, well-mannered, polite and was a very inventive blues guitar player. I consider myself lucky to have been able to play with Luther. When I'm playing guitar, he appears in my mind on many songs.

Jack Bonus

In 1977, while I was playing a big-band show, in walks Jack Bonus pacing back and forth like a caged lion. I found out right away he was somewhat inept socially, as so many frustrated musicians are. He was living in his 1972 Ford Pinto station

l-r Barry Lowenthal, Scott Urquhart, Walter Dickson, Jack Bonus and Bruce Gordon, 1990 wagon and needed a place to live. Soon after we played some music gigs, I moved him into my basement. I found his flute and saxophone playing imaginative, intense, and he played with fire. He put out one self-titled record in 1972 with Grunt Records, featuring Freddie Roulett who also played with Sky Blue Band.

Charley Spivac

This trumpet player is best remembered from the big band era of the late 1930s through the 1950s. He was known as "the sweetest trumpet in the world." Having been a big band leader, he had plenty of advice for me. He was on the road at the time with Woody Herman and talked of Gene Krupa being one of his childhood friends to Tommy and Jimmy Dorsey. I would say he was a bit manic. When in town, he stayed at my house on the couch and we enjoyed each other's company. One night after finishing *Perdito* at Specks, Charley borrowed Ned's trumpet and began playing Stardust with two trumpets in perfect harmony. After he finished, everyone was quiet for a few moments, then began clapping hands in amazement. It was one of those nights that affirmed my belief in God. Charley played and recorded with the top swing bands. I was amazed at how many records Charlie Spivac and his orchestra released.

Charlie said the movie, *Young Man with a Horn* (1950), was based on his character.

Terry Haggerty

Back in 1969, I had the pleasure of being at a live outdoor concert of The Sons of Champlin in Fairfax, CA. Every one of my guitar-playing friends was talking about Terry Haggerty. There he was, rocked back, playing what looked like to be an L5 Gibson through four Fender Twin Reverbs. At that point, I was playing through a Silvertone. I climbed the fence to get a better look. Right then they played a song called *Get High*. The place went crazy! Terry's been a great mode of inspiration and we still play on the Fourth of July. Believe me, that show was one of those moments.

Terry Haggerty and the Sky Blue Band, 1991
l-r Clem Healy, Terry Haggerty, Barry Lowenthal, Jack Bonus, Walter Dickson, Scott Urquhart

Harold Jones

In 1982, I bumped into a good friend of mine, Denise Jones. She tells me she's with one of the greatest drummers in the world. His name is Harold Jones, she says, "I call him Jonesy." That was it. When I heard he played with Count Basie, I really wanted to meet him. The list of people he has played with is almost forever: Count Basie, Sarah Vaughan, Ella Fitzgerald, Ray Charles, Tony Bennett, B.B. King, Lady GaGa, Willie Nelson, Natalie Cole and Jimmy Smith–the list goes on.

We started playing and recording in 1992 and to this day he has been one of the all-time great inspirations. It's his cooler than cool temperament that's been exactly what I love. To learn more about Harold, check out his biography, *Harold Jones: The Singer's Drummer*, by Gil Jacobs and Joe Agro.

One of Harold's philosophies is–the power of suggestion.

l-r Harold Jones, Terry Haggerty, David Russ, 2018 (Photo by Miriah Fox)

Jerry Garcia

Dickson Company HVAC, 1998.

When I was in the fifth grade in 1965, I worked weekends at my father's sheet metal shop at 708 West Francisco Blvd. in San Rafael. I was sweeping up and heard some music playing nearby. I got on my bicycle and headed toward the freeway overpass. There were a bunch of young musicians playing music in a warehouse right under the freeway. They were the Grateful Dead. Some of the musicians tried to kick me out but Jerry would always say, "Let him hang out." Jerry, always the nice guy. With his precise melodic playing and his placement between rhythm and lead guitar, he always seemed to me of having the feeling of an expression of genius.

I was lucky to pickup one of his guitars from the 1970s era. Jerry used this guitar on the 1971 double-album, *Grateful Dead.*

It's impossible for me not to be influenced by this great guitar player.

You can find pictures online from Grateful Dead, April 5th, 1971, Manhattan Center, NY.

I traded a 12-string guitar for this guitar, back in about 1979. This guitar was brought to me with the neck broken off at the body.

1961 Gibson Les Paul Custom SG-style heavily modified by Rick Turner then sold to Jerry Garcia around 1970-71 (Photo by Frank Lewis) Garcia-Turner Peanut Guitar

Klaus Kinski

Klaus Kinski wasn't a musician but a movie star who moved nearby. He was another genius who seemed to have a type of manic disorder. His advice to me was always the same, "don't quit your job" with his accent. He absolutely hated the business of the arts and pointed out the music business was one of the worst.

He would stop by the studio or invite me up to his house and we would drink a few of his favorite beers. If he was in the mood, he'd snap in his vampire fangs and act out. What a character with his wild, silver hair blowing around in his open jeep.

William "Willum" Schuler

During the exploding music scene around the Bay Area, Willum and other friends decided to leave Shreveport, LA, to try the dream of making it in the music business. In 1968, he made it to Forest Knolls and we met at an outdoor acoustic guitar jam when I was 14 years old. We became friends and he was amazed that we had a practicing band at the Dickson Brother's garage in Forest Knolls.

Harold Jones, William Schuler and Bruce Gapinski who has been with Sky Blue Band on and off since 1977 (Photo by Frank Lewis)

Willum had simplicity and a down home feel with his Southern droll and his country blues style. I helped produce one album for Willum around 1998. Later on, Dan Healy came down to finish the CD and it's one of my favorites. Dan did a great job recording a few more songs of Willum's and mixing out the album.

All original songs by William Schuler recorded with Sky Blue Band

Dave Alexander

Dave Alexander, also known as Omar Sharriff, was a blues piano player working around the Bay Area from about 1957 on. In the 1970s, a friend of ours was bartending a club on Fillmore Street, San Francisco. He invited us over to hear Dave's band. This started a great listening period for me since I love trios. Another person in the band was Eddie Adams the bass player. Eddie was one of the most solid bass players I had ever heard. Dave's trio taught me a lot about the down-to-earth feeling of the blues.

Bert Bales

When I became 21-years-old, the band got invited to a club in San Francisco called, Dick's at the Beach, off of Highway 1. Bert Bales was in charge of the jam sessions on Sundays with a Dixieland rhythm section. He never said a word, he would play an intro on the piano to a song and the whole band would jump in and start jamming. Being one of the youngest members there gave me a feeling of importance. Dixieland music has a way of weaving and blending instruments that form a gumbo of expression. At any rate, I had a great time with Bert and the boys.

My Parents

Thelma and Bill Dickson, 1940

My Mother, Thelma, studied painting, sculpture, loved watercolors and was fascinated with mosaics. At age 45, she became interested in opera and began taking singing lessons. I watched her study music notation on the piano and foreign languages. She was appointed the president of the Marin chapter of the United Nations and took me to some of the big meetings held in San Francisco. Around 1971, she was on a United Nations World Tour and was jailed in Vietnam as a spy. She was released 2-3 weeks later and recovered from dysentery in a hospital in Italy, then fully recovered in England. She continued her music with the help of Nancy Bennett and Phil Wood until she was in her early 90s and later died at home in 2017 at the age of 98.

My father, Alexander "Bill" Melville Dickson (grandson of Robert Dollar), was a steady man with a great sense of humor. He traveled as a young man on the Dollar Steamship Lines and learned the operation of traveling. He then served in the Merchant Marine in World War II. His main passion in life was flying and we had many adventures from Canada to Mexico with a few close calls. I got to experience weightlessness to diving down on a three-point hydro race boat at Tahoe called Miss Bardohl in a single engine retractable aircraft, called a Swift. When the pilot of the boat looked back and saw us coming down at him, he simply floored it and produced a 50-foot water rooster tail and left us behind. Never a dull moment with Dad. (Search on the internet, ride along with Miss Bardohl.)

Thelma Dickson and Nancy Bennett listen back to a track of theirs recorded by Dan Healy. (Photo by Frank Lewis)

Thelma and Walter being creative (Photo by Frank Lewis)

My Dad and Uncle Bill

Dad and my uncle on my Mom's side, Bill Simpson, had a piper cub single-engine airplane. One morning, they came up with a crazy idea. They took off in San Rafael and headed toward San Francisco looking for a ship. There it was, a large aircraft carrier. As they got closer, they saw there was no air traffic around it and the deck was clear. So, there they went down, down, down and actually landed the piper on its deck. Well, the Navy didn't like that and they were interrogated all day. Finally, late in the day, the commanders allowed my dad and uncle Bill to take off and fly away.

There were quite a few stories like that but bless them both. My Uncle Bill was the second best shot in Patton's 3rd Army in World War II. He and his sergeant were at the point position of the 12th Armored Division nicknamed the "Suicide Division" by the German Army. In a movie, he is portrayed as a private guarding the first concentration camp that the Americans captured. Being the lead scouts for the 3rd Army, Uncle Bill and his sergeant were the first Americans to come across a concentration camp. He shot around the main lock on the gate and kicked the gate open for the officers to inspect. His reaction, "horrific". It wasn't long after then that he was wounded in action but not before he shot down a Messerschmitt 109 with his 30.06 rifle. He and his sergeant were able to capture the pilot who was a colonel.

Miriah Fox

I'd like to thank Miriah Fox and her family, the Foxes, for their love and support. Thank you for your patience and your love. One can learn from Miriah's grandmother's book, *Beautiful Soul,* by Bella Kurant Fox Slamovich.

The Depth of Thinking

My hope is to bring down the cost of medical treatment with a more positive approach in the understanding of molecular and cellular vibration through the magnetic field. My thought is to treat disease at home through the possibility of reverse or harmonic frequency applied to the areas of needed repair. Through experiments, it shouldn't take long to find out if a disease responds to a logical frequency evaluation once the key root is calculated.

I believe there is intelligence in the Earth's magnetic field, being influenced by the Sun's magnetic field, being influenced by the galaxy and then by the universe.

Aneice Taylor, founder of In Spirit which serves people paralyzed from neurological disorders. Aneice and Walt have a long, interesting relationship.
In loving memory
October 22, 1944 - June 16, 2020

Walter Dickson at the Papermill Creek Saloon, Forest Knolls, CA (Photo by Frank Lewis)

The photo above is the interior of Speck's Irish Coffee Bar circa 1954. In the center-left, in the dark hat, is my Uncle Bob Simpson. The bartender is Speck McAuliffe. On my 28th birthday, I was partying at Speck's til about 4 am. About an hour later, the bar burned to the ground. This is the bar that Sky Blue Band played most Friday nights during the 1970s and early 1980s.

One of the main differences in a family-owned bar compared to a corporate bar is the feeling of being included in the family. As time went on, Speck became one of the band's mentors. One of his famous sayings was, "It's nice to be nice".

Permission to use above poster granted by Barbara Perkins

In remeberance of Bill Vitt

(May 6, 1943-July 16, 2019)

He was truly an iconic inspiration.

Bill was a drummer with the Jerry Garcia Band. He was a studio drummer working the West coast with Eydie Gorme, Freddie King, Mike Bloomfield, Sons of Champlin, Sky Blue Band and a host of others.

l-r Bill Vitt, Walter Dickson (Photo by Phil Wood)

l-r Hadi Al' Saadoon, Dave Farey, Steve Schuster (Photo by Joy Saadoon)

Frank Lewis. (Photo by Walter Dickson)

Michael McQuilkin, keybaords

Wailing Sam Carleton, blues harp, vocals

Pete Lind, drums

Peter Penhallow, keyboards

Jerry Marquez, drums

Liam Hanrahan, bass

Acknowledgments

I'd like to thank my family for their support Grace, Mel, Dan and my cousin Bernie Simpson.

Drummers: Harold Jones, Barry Lowenthal, Gary Silva, Sam Schaffer, Brian Fischer, Pete McConnell and always the lovely Diane Vitalich. Special thanks to Bill Vitt, Greg Ellis, Howie Cort, Bobby Rydel, Michael Carabello and Chris Jennings.

Horn players: Darrel Parker, Dale Alstrum, Tim Cain, Rita Thies, Dave Farey, John Farey, Steve Schuster, Snooky Flowers, Hadi Al' Saadoon and Kenny Jacobs.

Bass players: Scott Urquhart, Bruce Gapinski, David Russ, Bill Everett, Bobby Vega, Mario Cipollina, Mike Shae, John Beyers, Clyde Niesen, and Frank Lewis.

Guitar players: Terry Haggerty, Elvin Bishop, Mike Manning, Joe New, Warner Bayley, Kevin McConnell, Trygve Tryggestad, Archie Williams, Chris Hayes, Tyrone Dorian, Shannon Daly, Jeremy D'Antonio and Darren Nelson.

Keyboard: The Wonderful Marla Hunt Hanson, Phil Wood, Bruce Gordon, Nick Sutton, Dan Dickson, Michael McQuilkin, Howard Wales, Nancy Lyle Bennett, John Allair, Mike Duke, Dan Shea and Mike Lipskin.

Advisors and Friends: Evy McPherson (muse), Lisa Kindred, Aneice Taylor (www.inspirit-marin.org), Deb Cooley, Geoff Davis, Lowell Hendrickson, Joe Hopkins, Tyrone White, Wayne Canney, Katie Guthorn, Kira Thelin, Bob Brown, Angela Straily, Lorry Azevedo (Hat Racing Enterprises), Sam Watson (aka Wailin' Sam), Alexander McQuilkin and the San Geronimo Valley Stewards.

Special love to the Three Runkeys.

Tech Support: Lynn and Jim Archer, Dan Healy, David Trouse, Randy Smith (Mesa Boogie), Jay Daniels (Black Cat Studio), Judy Kaufman & John Pederson (Amazing Grace Music), Courtney Toriumi, Jared Litwin (Papermill Creek Saloon), the always sensitive Benny Shea, and the wonderful Mike Bendinelli (Mesa Boogie).

l-r Dan Dickson, Walter Dickson, Randy Smith, Fae McNally, Bruno Blunk, Bill Everett, Nick Sutton, Kevin McConnell, Greg Longfellow. Sky Blue Band playing at a party, circa 1975.

Notes

1. The Developmental Clock of Dental summarized from the internet by searching PMCID; PMC2667916. The development of children's teeth can be measured in 24-hour cycles as the layering processes of enamel counted as cross striations up to approximately six years of age. This article goes deep into the study and talks of circadian rhythms. Through a microscope looking at this layering process enables the exact count of 24-hour periods that dates to prenatal life of the individual. This counting of time relates to my statement "not that night or day changes the magnetic frequency" (page iii).

2. A Tuning Experiment. My thought was to take a grand piano into a hyperbolic chamber and conduct tests in + and – atmospheric conditions. The idea is to find where the temperament is most likely as close to perfect under these conditions as possible. The only way to scientifically test is with recording equipment and frequency monitoring devices plus a world class piano tuner. This might answer one of my questions about the Pythagorean System and temperament.

3. Many families of insects and animals live in different categories of the same frequencies generated by the magnetic field. From the four major frequencies I mentioned in the book comes octave divisions which relates to the sizes of the insects and animals. I first noticed this when conducting tests on flies. The size of two types of flies seem to vary according to which octave they related to. My point is the relation of magnetic frequency divides life into categories of size. This may be important for the study

of viruses and microscopic biology.

4. When mentioning the low note of the conversation, this statement means in general conversation. I have found that when people completely relax during conversation, usually at the end of a phrase, the note can drift down three half steps which equals the mystery frequency. In other words, if the high note of the conversation is A (equals low frequency) and F# being three half steps below the A (equals high frequncy), then the mystery frequency becomes D# which makes a diminished chord.

5. My explanation on the subject of energy changing frequencies (page 23) all started in California with me listening to live TV in the A.M and early P.M. from New York. As the planet travels counterclockwise (looking down from the north) and rounds into darkness, I was no longer able to sense magnetic frequency pitch changes traveling east to west. When New York passed into darkness, something changed.

When the Sun's magnetic energy first contacts the trailing edge on our planet's system, the Sun's magnetic field can change Earth's magnetic frequency at that location. From the trailing edge of Earth's magnetic field, the Sun's electric energy envelopes and circles our planet toward the leading edge of travel around the sun.

The Earth's magnetic field that faces the sun is compressed by the Sun's plasma energy as it blows past earth. Our compressed magnetic energy facing the sun also creates a decompressed zone of magnetic energy behind earth's dark side. The Sun's magnetic energy causes energy to travel a shorter distance in the compressed zone versus the decompressed zone being a more open and relaxed environment.

The two energies traveling at two different distances and possibly at two different energies meet at the approximate position of 2 a.m. to 4 a.m. causing the 24-hour circadian rhythms. I realized the magnetic field stays basically static as the earth rotates inside its field. The 4 a.m. area of the magnetic field conacts Earth moving counterclockwise looking down from the North.

West Marin, CA – an environmental analogy. Endangered species, Coho salmon and steelhead trout. A lesson in political science.

To figure out a problem, it becomes necessary to understand how the problem developed. When studying the tuning system, I read history and listened closely to the environment.

In West Marin, California, we have an out of tune situation with our community. Reading and listening to people, experts, politicians and special interest groups, here is how it boils down.

The creek system in San Geronimo Valley.

1. Years ago, there was a chemical spill from our local water treatment plant in Woodacre, California. This chemical, I believe, is called alum or aluminum sulphate. It is used in the process of separating particulates in drinking water. The chemical spill killed the Coho and steelhead and other life from the plant itself on the upper end of the creek system in Woodacre all the way down to parts of Lagunitas, approximately 3 miles

away. When talking to the community about writing this synopsis, I was told there have been other chemical spills as well.

2. The next disaster that happened to our creek system was a fire at the San Geronimo Valley Community Center in a building that housed an environmental special interest group teaching children about our local fish and animal environment. This building was approximately 5-10 feet from a tributary that led to the main San Geronimo creek system, approximately 1 mile downstream from our water treatment plant ran by Marin Municipal Water District (MMWD). When the fire department showed up to extinguish the fire, the order was given to use foam instead of water as a training exercise. The result was the creek was chemically contaminated and killed most life forms from San Geronimo to parts beyond Lagunitas.

Another interesting thing about this fire was the foam contained a chemical called alum or aluminum sulphate. The other fact that was overlooked in the investigation was how the fire was started.

I had just rebuilt the building's direct vent propane furnace to exact factory specifications. A week later, I did a second inspection on the building because of the fact that school children were frequently there. About a week later, the building burned to the ground. The next morning after the fire I stopped and witnessed a fireman washing the foam into the tributary that leads to the main creek system. Two days later, I drove to the fire department and told them I was the last technician to work on the building as I had just rebuilt the furnace system. They quickly and nervously said the case was closed and blamed the portable electric heater for the fire. I found that interesting as the electric heater was no longer being used. Free-standing electric heaters generally have a fail-safe device that turns the unit off if it falls over. This leaves me with the feeling that the building may have been intentionally set on fire.

3. I received a phone call that MMWD was going to do a fish count of Coho and steelhead in parts of Woodacre, including my family's ranch. I later saw four people outfitted in expensive water gear coming up the creek. One man was wearing a large backpack with a 36-42" wand in his right hand. They were friendly and I joined them and watched how the count was done. They were walking up the creek counting the fish from pond to pond. Their method was simple. They inserted a wand that was connected to a battery in his backpack similar to a cattle prod. Everyone stood back as the man pulled the trigger and electrocuted the pond causing the water to stand up about ¾ of an inch in a fierce and jolting vibration. Soon after all of the fish floated to the surface. One lady netted the fish and the other separated the fish into two buckets, one for the living and one for the dead. The ratio was about 30-35% killed and 65-70% alive. They measured the fish and wrote the numbers in a book and repeated this process to the top of the northern Woodacre creek system. They took the dead fish to their vehicle down the road. The size of the minnows was from 2-3" long. Later I called a California Wildlife official in Sonoma County and asked about this procedure. It was explained that MMWD had a permit to kill 15% of the fish.

There's another kind of fish counting system that was used and I participated in observing the methods used of counting and moving fish in the mid to late summer. The method used were handheld nets. This procedure was simply dipping the nets into the ponds and scooping up as many fish as they could. Some fish tried desparately to evade the nets and consequently were injured or killed in this process. This procedure was done year after year by volunteers. The fish that were captured were put into one bucket. After the counting of the fish, they were eventualy moved downstream to bigger ponds. I'm going to briefly mention two problems in this process-1: the injury and deaths of the fish not to mention the half-full, five gallon buckets were insufficient in size which caused more fatalities because of lack of oxygen; 2: over the course of moving the fish to bigger ponds downstream, the overpopulation of the fish caused starvation. After witnessing this for 3-4 years in a row, I contacted Trout Unlimited. They sent one of their representatives over and we did a creek walk. He explained that the procedure was mishandled and they should have used at least a 10 gallon buckets for fish transportation. Later on, I met the president of Trout Unlimited.

4. Back in about 2010-12, I met with some Lagunitas residents that I had been to many meetings with. They explained a situation with a private environmental corporation on their private property. This was before people in this area knew anything of property rights to do with the creek systems on their property. The corporation had been setting up nets crossing the creek during the main months of the spawning period. The nets were

placed east of Peter's Spillway (Kent Lake) in Lagunitas which is where the San Geronimo Valley community starts.

The nets stretched the entire width of the creek system which caught the egg-laden Coho and steelhead on their way to the upper creek system. I was told that they caught so many fish that they removed them by pickup truck. I realize that this was secondhand information but later I received a phone call and was told to walk out on the Spring Road Bridge in Lagunitas off of Sir Francis Drake Blvd. There I witnessed a man manipulating a fishnet about 3' high and 35' long just west of the bridge. Later on, I wrote a letter to the head of California Fish and Game Commission about the last two incidents with no reply. By then, our fish population had dropped by approximately 70%. All of this activity contributed to what led to our fish population being put on the endangered species list. The question of oceanic environmental conditions played no significant part of our fish population decreasing. I find it interesting that my calculation of contaminated water from our community damaging our creek system is approximately 5-8%.

Peter's Dam (Kent Lake) also has an interesting history as in expanding its capacity in the late 1970s to early 1980s. The raising of the dam resulted in less water flow and the depletion of the Coho and steelhead population.

My mind has a problem dealing with these facts as the counting process of the fish had a killing effect of over 4 times the amount of the estimated damage of contaminants released by our housing community.

What I've learned is, no community can understand or deal with the overwhelming laws and the environment blitz of finger pointing, guilt, lawsuits and confusion.

When peeling back the curtain of this Wizard of Oz nightmare, lies an overwhelming sense that our community has been used, abused and misled from the top to the bottom. The other fact is that 90% of descisions and laws are handled behind closed doors by special interest groups collecting millions of dollars in a play ball or get sued environment.

In participating at the larger environmental meetings, one of the first things that was said from our County Supervisor was, "the door to science is closed." By using this statement puts parameters in the pockets of the government and special interest groups and curtails the public from participation.

During this period, I participated in meetings with multiple state, federal and county officials. They used maps that excluded the two main dam systems located on the two major parts of the four branches of our creek system. Peter's Dam (Kent Lake) in Lagunitas and Nicasio Reservoir Dam were the two major factors of Coho and steelhead's loss of environment, not to mention the other dam systems that tie into the same environment.

Being the end of the summer of 2019 with millions of dollars of work being done in the local creek system, we are guaranteed that the Federal, State and County governments will take credit in the increase of Coho and steelhead population. My reasoning for thinking this is special interest groups won't use nets or use less of them. The counting of fish most likely will be modified so the populations can increase. MMWD most likely will stop poisoning the creek adding to the fish population. The Marin County Fire Department most likely won't use foam on fires near our creek system or be flushing their fire hydrants near our creek which will result in the expansion of fish population. If these entities stop attacking our creek system, the fish population is guaranteed to increase.

The Final Argument

Special interest groups have the ability to sue MMWD over the fact of destroying approximately one–third of the Coho and Steelhead habitats. With this hanging over Marin County's head lies the ability of the parties to get exactly what they want in legal fees and lawsuits. In other words, they can hardly lose a case.

Until Marin County has legal proof that the dam systems can't be sued, we as citizens are at a disadvantage. The fact is, the argument can be made that the dam system weakened the species as a whole in their ability to survive is inarguable.

It's not that the scientists are lying, it's just that they tell a partial truth as a one size fits all that allows their bag of politcal power to control the public.

The laws passed here in Marin County California will be the model for the West Coast and then for the entire United States.

Public Land Trust Purchases San Geronimo Golf Course

I came across a Point Reyes Light newspaper dated September 5, 2019. On the front page it says, "State bill reacts to golf course ruling." A.B. 782 was signed into law August 30, 2019, in Sacramento apparently allowing public agencies to be exempt from conducting the California Environmental Quality Act (CEQA) reports on any property they want to purchace, depending on if it is in the category of sensitive or threatened land.

As we are winding up the third edition of this book, I find it hard to stay out of politics. Unfortunately people out of the special interest groups like myself, find it very discouraging and leaves me with the feeling of negativity. Right now in West Marin, the answer to the environments of the creeks rehabilitation amounts to the size of the portfolio of special interest groups including tractors in the creek and, unfortunately, a lot of debatable science.

My prediction of the fish population increase because of less interference of special interest groups will bring our new govenor, Gavin Newsome, out to West Marin for a photo shoot to promote his 2028 run for presidency. Of course, all the special interest groups will be prospering as the laws applied here will be swept across the country.

October 2, 2019

Yesterday, I was home recovering from a neck injury and writing in my notebook. I noticed when I breathed in it was the key of A# and when I breathed out it was the key of G. These two keys were the frequencies of the magnetic field as my notebook reveals. I've known this for years and years and just wanted to include it.

Recently the magnetic field has moved one-half step up to the key of G/A#.

My thought of the expansion of the second edition of this book is, in my belief in nature, there may be a reason for everything.

This spider web is similar to the graph on page 50.
The premise of this book was asking myself the question - where does life get its information?

One of my favorite memories of being young was the many hours I sat watching a spider spin its web. As David and I finish the second edition of this book, I noticed this spider web outside the studio door.

The question becomes where does the spider get its information as how to construct its web. The same question is applied to the birds, flies, horses, dogs, humans, trees, etc.

I heard a story from a friend that one of his friends committed suicide over having tinnitus. Being diagnosed with tinnitus at a very young age, left me with a feeling of misunderstanding. Not only was it irratating and effecting my personality in a negative way, all the people that I talked to who had tinnitus had no answer for what it was.

It wasn't until the mid-1990s that I learned to study and live with the understanding that I was determined to find out what it was and where it came from. I'm hoping that this book opens up peoples ears to understand that it is just part of the magnetic electronic nature of life.

Understanding what I hear has helped me with all of my relationships with people, friends, animals, and has given me insight to write this book in a positive light.

Walter Dickson

SKY
BLUE
BAND

skyblueband@comcast.net

The Thought of Thinking Common Sense

The rise of carbon as parts per million in itself is a measurement of heat as an exchange in chemical energy or exothermics creating carbon gas and other chemicals.

I've spent decades working on gas and electric furnaces, water heaters, boilers, appliances, and air conditioners. Air conditioners and refrigeration systems create heat by compressing the gas into a liquid before the evaporation process begins. The evaporation of this compressed liquid creates cooling at a much less efficiency level. The world is being heated, not so much from greenhouse gases, but by the chemical reaction of combusting fuel. This creates the rise of parts per million of carbon. Carbon dioxide and carbon monoxide, as gases, are the byproducts of burning, creating heat.

For instance, all new and older car engines run at approximately 195 degrees Fahrenheit and dissipates this heat through radiators to the cooler atmosphere. This is caused by 195-degree thermostats that are built into the engines. Imagine lowering the thermostat temperature to 145 degrees cooling the radiators, engines and automatic transmissions down 50 degrees. This would drop the heat output by approximately 25-30% affecting over 270 million registered vehicles in the United States alone. I'm aware that this may change the combustion temperature and alterations might be considered for vehicle emissions control.

The information in this book is still being modified as I am working on other books all the time.

Made in the USA
Monee, IL
05 November 2026

67953747R00074